Illustrator:
Agi Palinay

Editors:
Evan D. Forbes, M.S. Ed.
Walter Kelly, M.A.
Jo Ann Merrell

Senior Editor:
Sharon Coan, M.S. Ed.

Creative Direction:
Elayne Roberts

Product Manager:
Phil Garcia

Imaging:
Alfred Lau

Research and Contributions:
Bobbie Johnson

Photo Cover Credit:
Images ©PhotoDisc, Inc. 1994

Publishers:
Rachelle Cracchiolo, M.S. Ed.
Mary Dupuy Smith, M.S. Ed.

Hands-On Minds-On Science

Magnetism and Electricity

Intermediate

Author:
Mel Feigen, M.A.

Teacher Created Materials, Inc.
6421 Industry Way
Westminster, CA 92683
www.teachercreated.com

©1994 Teacher Created Materials, Inc.
Reprinted, 1999
Made in U.S.A.
ISBN-1-55734-646-1

The classroom teacher may reproduce copies of materials in this book for classroom use only. The reproduction of any part for an entire school or school system is strictly prohibited. No part of this publication may be transmitted, stored, or recorded in any form without written permission from the publisher.

Table of Contents

Table of Contents *(cont.)*

How Do Electricity and Magnetism Work Together?

Station-to-Station Activities

Management Tools

Glossary

Introduction

What Is Science?

What is science to children? Is it something that they know is part of their world? Is it a textbook in the classroom? Is it a tadpole changing into a frog? A sprouting seed, a rainy day, a boiling pot, a turning wheel, a pretty rock, or a moonlit sky? Is science fun and filled with wonder and meaning? What is science to children?

Science offers you and your eager students opportunities to explore the world around you and to make connections between the things you experience. The world becomes your classroom, and you, the teacher, a guide.

Science can, and should, fill children with wonder. It should cause them to be filled with questions and the desire to discover the answers to their questions. And, once they have discovered answers, they should be actively seeking new questions to answer. The books in this series give you and the students in your classroom the opportunity to learn from the whole of your experience—the sights, sounds, smells, tastes, and touches, as well as what you read, write about, and do. This whole-science approach allows you to experience and understand your world as you explore science concepts and skills together.

What Are Magentism and Electricity?

Magnetism and electricity are natural energy forms that have existed since the beginning of time. Although these energy forms have been around for many years, they have only had practical use in our society for the last several hundred years. We cannot see, hear, or smell the force of electricity or magnetism. What we know about these energy forms is by what they do.

Lightning was one of the first forms of electricity known to humans. Later, scientists generated electrical charges (static electricity) by rubbing certain materials together and, with the creation of batteries, learned to store electricity. Electricity is in the air, in animals, and even in our human bodies.

Magnetism is a significant force in nature that can make some things move towards each other, move away from each other, or even stay in one place. Surrounding the earth is a great magnetic field exerting this force. Just as magnetic fields lie around the earth, they also lie around every magnet. Magnetism is an essential force that allows us to do many important things we do with electricity.

Electricity and magnetism are two very useful, powerful forces. The knowledge of these forces collected over the years has permitted us to create amazing things, from heavy duty motors to micro-computers, from telephones to space communications. Safety and conservation should always be a top priority whenever manipulating either of these two great forces.

The Scientific Method

The "scientific method" is one of several creative and systematic processes for proving or disproving a given question, following an observation. When the "scientific method" is used in the classroom, a basic set of guiding principles and procedures is followed in order to answer a question. However, real world science is often not as rigid as the "scientific method" would have us believe.

This systematic method of problem solving will be described in the paragraphs that follow.

1 Make an OBSERVATION.

The teacher presents a situation, gives a demonstration, or reads background material that interests students and prompts them to ask questions. Or students can make observations and generate questions on their own as they study a topic.

Example: Show a magnet picking up an object through a piece of cardboard.

2 Select a QUESTION to investigate.

In order for students to select a question for a scientific investigation, they will have to consider the materials they have or can get, as well as the resources (books, magazines, people, etc.) actually available to them. You can help them make an inventory of their materials and resources, either individually or as a group.

Tell students that in order to successfully investigate the questions they have selected, they must be very clear about what they are asking. Discuss effective questions with your students. Depending upon their level, simplify the question or make it more specific.

Example: Can magnetic forces go through objects?

3 Make a PREDICTION (Hypothesis).

Explain to students that a hypothesis is a good guess about what the answer to a question will probably be. But they do not want to make just any arbitrary guess. Encourage students to predict what they think will happen and why.

In order to formulate a hypothesis, students may have to gather more information through research.

Have students practice making hypotheses with questions you give them. Tell them to pretend they have already done their research. You want them to write each hypothesis so it follows these rules:

1. It is to the point.
2. It tells what will happen, based on what the question asks.
3. It follows the subject/verb relationship of the question.

Example: I think magnetic forces can go through some objects, depending on their thickness.

The Scientific Method *(cont.)*

 4 Develop a **PROCEDURE** to test the hypothesis.

The first thing students must do in developing a procedure (the test plan) is to determine the materials they will need.

They must state exactly what needs to be done in step-by-step order. If they do not place their directions in the right order, or if they leave out a step, it becomes difficult for someone else to follow their directions. A scientist never knows when other scientists will want to try the same experiment to see if they end up with the same results!

Example: We will test several objects to see if the magnetic force goes through them.

 5 Record the **RESULTS** of the investigation in written and picture form.

The results (data collected) of a scientific investigation are usually expressed two ways—in written form and in picture form. Both are summary statements. The written form reports the results with words. The picture form (often a chart or graph) reports the results so the information can be understood at a glance.

Example: The results of the investigation can be recorded on a data-capture sheet provided (page 22).

6 State a **CONCLUSION** that tells what the results of the investigation mean.

The conclusion is a statement which tells the outcome of the investigation. It is drawn after the student has studied the results of the experiment, and it interprets the results in relation to the stated hypothesis. A conclusion statement may read something like either of the following: "The results show that the hypothesis is supported," or "The results show that the hypothesis is not supported." Then restate the hypothesis if it was supported or revise it if it was not supported.

Example: The hypothesis that stated "magnetic force can go through some objects, depending on their thickness" is supported (or *not supported*).

 7 Record **QUESTIONS, OBSERVATIONS,** and **SUGGESTIONS** for future investigations.

Students should be encouraged to reflect on the investigations that they complete. These reflections, like those of professional scientists, may produce questions that will lead to further investigations.

Example: What kinds of objects will magnetic force not pass through?

Science-Process Skills

Even the youngest students blossom in their ability to make sense out of their world and succeed in scientific investigations when they learn and use the science-process skills. These are the tools that help children think and act like professional scientists.

The first five process skills on the list below are the ones that should be emphasized with young children, but all of the skills will be utilized by anyone who is involved in scientific study.

Observing

It is through the process of observation that all information is acquired. That makes this skill the most fundamental of all the process skills. Children have been making observations all their lives, but they need to be made aware of how they can use their senses and prior knowledge to gain as much information as possible from each experience. Teachers can develop this skill in children by asking questions and making statements that encourage precise observations.

Communicating

Humans have developed the ability to use language and symbols which allow them to communicate not only in the "here and now" but also over time and space as well. The accumulation of knowledge in science, as in other fields, is due to this process skill. Even young children should be able to understand the importance of researching others' communications about science and the importance of communicating their own findings in ways that are understandable and useful to others. The magnetism and electricity journal and the data-capture sheets used in this book are two ways to develop this skill.

Comparing

Once observation skills are heightened, students should begin to notice the relationships between things that they are observing. *Comparing* means noticing similarities and differences. By asking how things are alike and different or which is smaller or larger, teachers will encourage children to develop their comparison skills.

Ordering

Other relationships that students should be encouraged to observe are the linear patterns of seriation (order along a continuum: e.g., rough to smooth, large to small, bright to dim, few to many) and sequence (order along a time line or cycle). By ranking graphs, time lines, cyclical and sequence drawings, and by putting many objects in order by a variety of properties, students will grow in their abilities to make precise observations about the order of nature.

Categorizing

When students group or classify objects or events according to logical rationale, they are using the process skill of categorizing. Students begin to use this skill when they group by a single property such as color. As they develop this skill, they will be attending to multiple properties in order to make categorizations; the animal classification system, for example, is one system students can categorize.

Science-Process Skills *(cont.)*

Relating

Relating, which is one of the higher-level process skills, requires student scientists to notice how objects and phenomena interact with one another and the change caused by these interactions. An obvious example of this is the study of chemical reactions.

Inferring

Not all phenomena are directly observable, because they are out of humankind's reach in terms of time, scale, and space. Some scientific knowledge must be logically inferred based on the data that is available. Much of the work of paleontologists, astronomers, and those studying the structure of matter is done by inference.

Applying

Even very young, budding scientists should begin to understand that people have used scientific knowledge in practical ways to change and improve the way we live. It is at this application level that science becomes meaningful for many students.

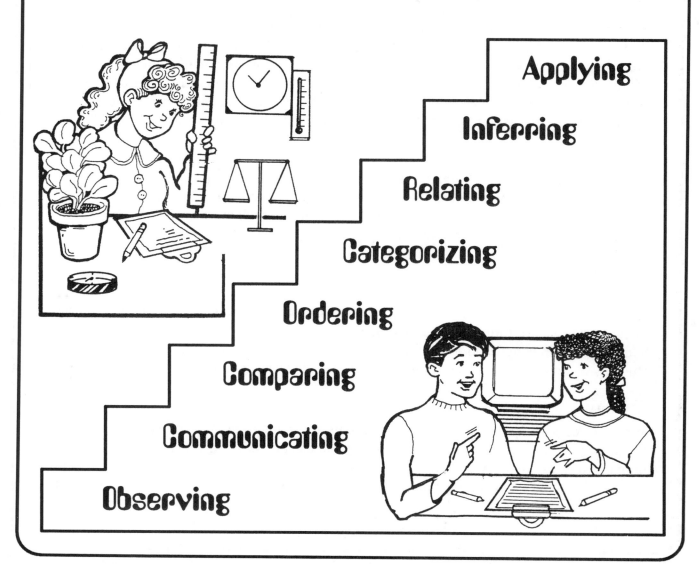

Organizing Your Unit

Designing a Science Lesson

In addition to the lessons presented in this unit, you will want to add lessons of your own, lessons that reflect the unique environment in which you live, as well as the interests of your students. When designing new lessons or revising old ones, try to include the following elements in your planning:

Question

Pose a question to your students that will guide them in the direction of the experience you wish to perform. Encourage all answers, but you want to lead the students towards the experiment you are going to be doing. Remember, there must be an observation before there can be a question. (Refer to The Scientific Method, pages 5-6.)

Setting the Stage

Prepare your students for the lesson. Brainstorm to find out what students already know. Have children review books to discover what is already known about the subject. Invite them to share what they have learned.

Materials Needed for Each Group or Individual

List the materials each group or individual will need for the investigation. Include a data-capture sheet when appropriate.

Procedure

Make sure students know the steps to take to complete the activity. Whenever possible, ask them to determine the procedure. Make use of assigned roles in group work. Create (or have your students create) a data-capture sheet. Ask yourself, "How will my students record and report what they have discovered? Will they tally, measure, draw, or make a checklist? Will they make a graph? Will they need to preserve specimens?" Let students record results orally, using a video or audio tape recorder. For written recording, encourage students to use a variety of paper supplies such as poster board or index cards. It is also important for students to keep a journal of their investigation activities. Journals can be made of lined and unlined paper. Students can design their own covers. The pages can be stapled or be put together with brads or spiral binding.

Extensions

Continue the success of the lesson. Consider which related skills or information you can tie into the lesson, like math, language arts skills, or something being learned in social studies. Make curriculum connections frequently and involve the students in making these connections. Extend the activity, whenever possible, to home investigations.

Closure

Encourage students to think about what they have learned and how the information connects to their own lives. Prepare magnetism and electricity journals using directions on page 82. Provide an ample supply of blank and lined pages for students to use as they complete the Closure activities. Allow time for students to record their thoughts and pictures in their journals.

The Big Why

The explanation behind the experience is provided.

Organizing Your Unit *(cont.)*

Structuring Student Groups for Scientific Investigations

Using cooperative learning strategies in conjunction with hands-on and discovery learning methods will benefit all the students taking part in the investigation.

Cooperative Learning Strategies

1. In cooperative learning, all group members need to work together to accomplish the task.
2. Cooperative learning groups should be heterogeneous.
3. Cooperative learning activities need to be designed so that each student contributes to the group and individual group members can be assessed on their performance.
4. Cooperative learning teams need to know the social as well as the academic objectives of a lesson.

Cooperative Learning Groups

Groups can be determined many ways for the scientific investigations in your class. Here is one way of forming groups that has proven to be successful in intermediate classrooms.

- **The Project Leader**—scientist in charge of reading directions and setting up equipment.
- **The Physicist**—scientist in charge of carrying out directions (can be more than one student).
- **The Stenographer**—scientist in charge of recording all of the information.
- **The Transcriber**—scientist who translates notes and communicates findings.

If the groups remain the same for more than one investigation, require each group to vary the people chosen for each job. All group members should get a chance to try each job at least once.

Using Centers for Scientific Investigations

Set up stations for each investigation. To accommodate several groups at a time, stations may be duplicated for the same investigation. Each station should contain directions for the activity, all necessary materials (or a list of materials for investigators to gather), a list of words (a word bank) which students may need for writing and speaking about the experience, and any data-capture sheets or needed materials for recording and reporting data and findings.

Model and demonstrate each of the activities for the whole group. Have directions at each station. During the modeling session, have a student read the directions aloud while the teacher carries out the activity. When all students understand what they must do, let small groups conduct the investigations at the centers. You may wish to have a few groups working at the centers while others are occupied with other activities. In this case, you will want to set up a rotation schedule so all groups have a chance to work at the centers.

Assign each team to a station, and after they complete the task described, help them rotate in a clockwise order to the other stations. If some groups finish earlier than others, be prepared with another unit-related activity to keep students focused on main concepts. After all rotations have been made by all groups, come together as a class to discuss what was learned.

Just the Facts

Prehistoric people were probably aware that certain rocks could attract or repel each other. These rocks are called *lodestones,* and they are considered natural magnets.

A magnet exerts a force that is invisible. You cannot see or feel it. However, when anything with iron is placed close to it, the magnet will attract it.

The word *magnet* comes from the ancient Greek town of Magnesia. Natural magnets were commonly found in this area of Greece.

Dr. William Gilbert was the first to show that the earth is a giant magnet.

Following are some facts about magnets:

- Objects attracted by a magnet contain iron or steel.
- Magnets have two unlike poles. They are given the names *north* and *south.*
- Unlike poles attract and like poles repel.
- The iron atom's magnetism comes from the electron.
- Magnetic force can pass through other solid objects.
- The earth itself is a huge magnet.
- Magnetism was first described by a Greek philosopher named Thales of Mellitus in about 500 B.C.

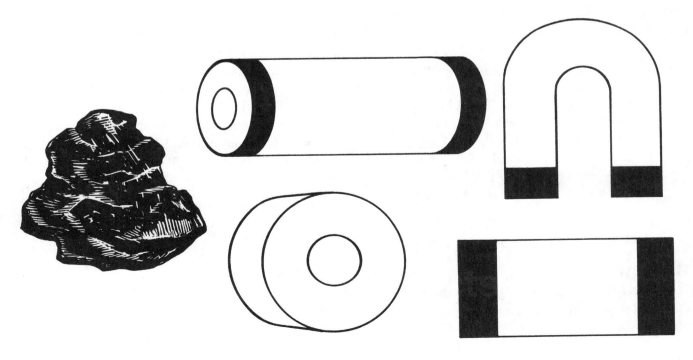

Great Separator

Question

How can you separate iron filings from salt?

Setting the Stage

- Have students try separating, by hand, grains of rice that have been mixed with salt or fine sand.
- Discuss with students the difficulties of separating by hand and brainstorm as to other methods that might work better.
- Discuss with students methods for separating other kinds of mixtures.

Materials Needed for Each Group

- iron filings
- a bar magnet
- several plastic bags
- plastic wrap
- salt or fine white, play sand
- small paper cup
- data-capture sheet (page 13), one per student

Procedure *(Student Instructions)*

1. Mix a bag of iron filings with sand or salt.
2. Wrap the magnet completely in a piece of plastic wrap.
3. Insert one end of the magnet into the mixture.
4. When the end of the magnet is covered with iron filings, hold the magnet over the cup and carefully unwrap it so that the filings fall into the cup.
5. Continue the process until all of the iron filings have been separated.

Extensions

- Have students repeat the experience, this time with crushed cereal. (Use one high in iron content.) What do you think the magnet picks up? How does our body use iron?
- Have students go out to the school yard with their magnets, making sure they are still covered in plastic wrap. Move the magnets around in the sand? What do you end up with?
- Have students fill a large cake pan with approximately 1" (2.5 cm) of distilled water. (Tap water will often contain iron.) Put it on a high spot where it will not be disturbed. (A flat roof works well if there is a safe way to reach it.) Leave it until the water evaporates. There should be some very small particles left. Run a magnet over these particles. Some of them should stick to the magnet. These are magnetic particles from meteorites (space dust, to be more exact) that have traveled through our atmosphere. This experience is best done in the spring and summer.

Closure

Have students complete their data-capture sheets and then add them to their magnetism and electricity journals. If your class went to the playground, have each group share experiences.

The Big Why

The iron filings are attracted to the magnet. The salt (or sand) is not influenced by the magnet and is left behind in the bag.

Great Separator *(cont.)*

Describe in a short paragraph what you observed and why it happened.

_____

_____

_____

_____

Would it work with either end of the magnet?

_____

_____

Why wasn't the salt picked up by the magnet?

_____

_____

You are building a processing plant that will separate recyclable items that have been collected from the trash. What are some methods of separation that you might want to consider for your plant?

_____

_____

_____

_____

_____

_____

_____

_____

_____

_____

Don't Give Me Any Static

Question

How do static charges act like magnets?

Setting the Stage

- Blow up a balloon and rub it against any material made of natural fibers. Gently place the balloon on the blackboard or wall. Then explain to students static forces will cause it to stick.
- Ask students what they know about static electricity. Encourage them to share their stories about being "shocked."

Materials Needed for Each Group

- salt
- pepper
- plastic comb
- piece of wool or flannel
- small clear dish
- data-capture sheet (page 16), one per student

Procedure *(Student Instructions)*

1. Place a small amount of salt and a small amount of pepper in the glass dish.
2. Rub the comb vigorously with the wool cloth.
3. Move the comb slowly to the salt and pepper mixture.
4. Move the comb even closer without touching the mixture. Finally, touch the comb to the mixture.
5. Record your observations on your data-capture sheet.
6. Test other items in the room to see what else is attracted by static forces.

Extensions

- Have students try this experience under different weather conditions. The amount of humidity in the air affects static electrical forces.
- Have students use different materials to find the best producers of static electricity.
- Have students measure the strength of static electrical forces by using the comb to pick up pieces of paper. Weigh the paper on a balance scale to find the strength of the force.

Closure

In their magnetism and electricity journals, have students write a paragraph comparing and contrasting magnetic forces and electrostatic forces.

Don't Give Me Any Static *(cont.)*

The Big Why

By rubbing the plastic comb with the wool or flannel cloth, it is charged with electrons. Free moving electrons from the material are transferred to the comb. The comb becomes negatively charged and acts similar to a magnet. The comb, however, does not have a north and south pole. The pepper rises first because it is lighter. If you come close enough, some of the salt will also jump to the comb.

The attraction is caused by the movement of electrons on the surface of the pepper. The negatively charged comb forces the electrons on the pepper to move away from the comb. The pepper becomes charged by induction; that is, it now has a negative side (away from the comb) and a positive side (closest to the comb). Since opposite charges attract, the pepper is drawn to the comb. Another way to separate the salt and pepper is to put them both in a small amount of water. The salt will sink to the bottom (it is heavier) and the pepper will float.

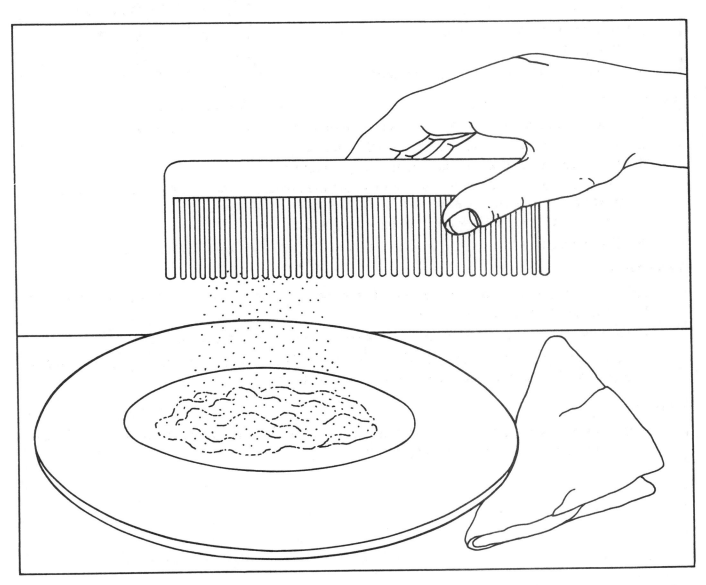

Don't Give Me Any Static (cont.)

Fill in the information needed.

Observations

What did you observe as the comb came close to the mixture? _____

_____

What happened when the comb was very close to the mixture? _____

_____

What happened when the comb touched the mixture? _____

_____

Conclusions

What causes the attraction between the comb and the mixture?_____

_____

Extension

How is the attraction that you observed with static electrical forces the same as it is with magnetic forces?

_____

_____

How is it different? _____

_____

Electrostatic Copy Machine

Question

How does a copy machine use electrostatic attraction?

Setting the Stage

- Have your class brainstorm any practical uses of electrostatic forces and static electricity.
- Explain to students that this activity is the basic principal behind the photocopy machine.

Materials Needed for Each Group

- a piece of wool cloth
- paper
- scissors
- some pepper
- a plastic petri dish and lid
- data-capture sheet (page 18), one per student

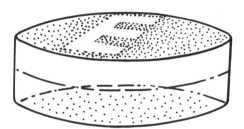

Procedure *(Student Instructions)*

1. Put some pepper into a petri dish and put the lid back on. Shake the petri dish so that the pepper is spread over the bottom.
2. Cut out a shape or letter from the paper that will fit on the petri dish.
3. With the wool cloth, rub the open area around the shape or letter on top of the petri dish. Hold the shape firmly so that it does not move. Rub 25 to 50 times.
4. Take the stencil off and flip the dish so the pepper now rests against the top. Flip the dish upright again.
5. Record your observations on your data-capture sheet.

Extensions

- Have students find out more about photocopying machines. Then have them write a one-page paper on what they learned.
- Discuss with students other ways things are copied.

Closure

Have students complete their data-capture sheets and add them to their magnetism and electricity journals.

The Big Why

Rubbing the plastic lid gives it a negative charge. That is, rubbing gives it a surplus of electrons. The pepper becomes charged by induction and is attracted to the negative areas. This is the same principal used for photocopy machines.

Electrostatic Copy Machine (cont.)

Fill in the information needed.

Observations

Draw a picture of the top of the petri dish after it was charged and then flipped over.

Where did the pepper stick? _____

_____

Why did the pepper stick to the rubbed areas? _____

_____

What charge did the plastic surface have? _____

_____

Explain how this principle can be used in a photocopy machine.

_____

_____

_____

_____

Wrapped Hand

Question

Is there a simple way to share with others the attraction of electrostatic forces?

Setting the Stage

Explain to your class that they have been asked to share a science experience and explanation with a younger class. (Make arrangements to do this.) Many students are familiar with magnets and being shocked by static electricity, but they are not familiar with the similarities between the two. This demonstration is simple to perform and a good illustration of the concepts.

Materials Needed for Each Group

- plastic wrap
- a wooden ruler
- a friend's hand
- data-capture sheet (page 20), one per student

Procedure *(Student Instructions)*

1. Hold a 2' (60 cm) piece of plastic wrap against the wall.
2. Rub the surface of the plastic wrap with your hand.
3. Drape it over the wooden ruler. The sides will repel each other.
4. Ask volunteers from the audience to put their hands between the plastic wrap. They will be amazed as the plastic wrap envelops their hands.
5. Once you have completed your classroom presentation, write it up on your data-capture sheet so it can be presented to another class.

Extensions

- Have each group practice their presentation before the class. Pick the best ones to present to a younger class.
- Have students repeat this experience at home for their parents.
- Have students try to find other experiences that seem to be magic tricks and put on a super science magic show.

Closure

Go to another class and have students take part in the experience. Students who are not involved can be part of the audience.

The Big Why

Rubbing the plastic wrap on the wall charged it negatively. When it was initially draped over the ruler, the two sides repelled each other (like charges). When someone inserted a hand, the plastic wrap moved the electrons on the surface of the hand (the like charges repel, causing electrons to move away from the area nearest the plastic wrap) until the hand showed a more positively charged side to the plastic wrap. Attraction of the two was the result.

Wrapped Hand *(cont.)*

Decide the parts each member of your group will play in the presentation of your experience. Write a script to be used during the presentation. You may make it seem like a magic show and add a lot of pizzazz. Your presentation should include the reason the plastic wrap separates when it hangs on the ruler and why it is attracted to your hand. You may use pictures in your explanation. Draw them on the back of this paper.

Title of your presentation: _____

Group members and their parts: _____

_____

_____

_____

_____

_____

Script to be performed:

_____

_____

_____

_____

_____

_____

_____

_____

Draw pictures to explain experience. (Use the back of this paper.)

Force Knows No Bounds

Question

Do magnetic forces go through objects?

Setting the Stage

- Have your class list as many kinds of forces and things that create forces as they can. Types of forces might include magnetic, electrostatic, and gravitational. Examples of things that create forces might include the wind, ocean waves, rocket engines, people, motors, etc.
- Discuss with students which forces might be able to go through objects.

Materials Needed for Each Group

- bar magnet
- several books of different thicknesses
- compass
- paper
- cloth
- aluminum foil
- various solid objects
- transparent tape
- iron nail
- data-capture sheet (page 22), one per student

Procedure *(Student Instructions)*

1. Stand a book up vertically. Put a compass on one side of it. Put the magnet on the opposite side of the book. Observe what happens. Record your observation on your data-capture sheet.
2. Try more books in between and observe the reactions on the compass.
3. Tape a piece of paper around the magnet. Try to pick up the nail with the magnet. Keep adding layers of paper around the magnet. Observe what happens. Record your observation on your data-capture sheet.
4. Complete your data-capture sheet.

Extensions

- Have students place two bar magnets together and repeat the experience. Have them record their results. Did two magnets change the results?
- Have students try placing different materials between the magnet and the compass. Have them record their results.

Closure

In their magnetism and electricity journals, have students write a statement about the magnetic field and its ability to permeate things.

The Big Why

Magnetic fields will pass through almost anything. The stronger the magnet, the greater it exhibits this phenomenon. It is similar to gravity in this respect.

Force Knows No Bounds *(cont.)*

Fill in the information needed.

Observations

Objects Used	No Change	Weaker Force	Magnetic Force	Completely Blocked
1.				
2.				
3.				
4.				
5.				
6.				
7.				
8.				

Write a general rule about the magnetic field and its ability to go through objects.

_____

_____

_____

_____

_____

_____

_____

Magnetic Induction, What's Your Function?

Question

Can we make magnets out of non-magnets?

Setting the Stage

- Ask your class what would happen if we cut a bar magnet in half. Would we still have a magnet?
- Ask your class what would happen if we keep cutting the magnet into smaller and smaller pieces. Would we reach a point when we no longer had a magnet?

Materials Needed for Each Group

- strong bar magnet
- pile of small nails or paper clips and various small objects
- long iron nail
- data-capture sheet (page 24), one per student

Procedure *(Student Instructions)*

1. Touch a long nail to a pile of paper clips or small nails. Observe and record what happens on your data-capture sheet.
2. Touch one of the poles of the bar magnet to the head of the nail. While keeping the bar magnet touching the head of the iron nail, place the pointed end of the long nail into the paper clips.
3. Move the bar magnet away from the end of the nail.
4. Use the bar magnet to stroke the nail (about 25 times). Always stroke in the same direction. Try picking up the paper clips with the nail after you have stroked it.
5. Complete your data-capture sheet.

Extensions

- Have students try to induce other objects to become magnets. Have students repeat this experience, this time with other types of magnets. Observe the number of paper clips they attract.
- Have students repeat step one of the experience ten times. Then have them average the number of paper clips picked up during the ten attempts.

Closure

Have students add their completed data-capture sheets to their magnetism and electricity journals.

The Big Why

Each tiny molecule of a metal acts like a magnet. The problem is that these pieces are all in a jumble, so that the whole piece does not act like a magnet. By stroking the nail with the magnet, the tiny atoms of iron line up into magnetic domains, much like students sitting in rows of desks. This makes the whole nail into a single magnet. A fancy name for this is magnetic induction. Over time, the iron atoms in the nail will become jumbled again due to natural atomic movements, heat (which increases the movement), or by jarring the nail by dropping it or hitting it. You will notice that your classroom magnets seem to become weaker each year that you use them. These magnets are doing the same thing, but more slowly. To preserve the strength of your magnets, always store them in a cool spot and place them together with south ends on top of north ends.

Magnetic Induction, What's Your Function? *(cont.)*

Draw a picture below of the pieces inside a nail as they normally are arranged, all in a jumble.

Draw a picture below of the pieces inside a nail after they are lined up into magnetic domains.

How does the nail behave when the iron atoms are all in a jumble? How does the nail behave when the iron atoms are all lined up? There are many materials that cannot be made into a magnet. Why do you think this is?

_____

_____

_____

Compass in a Bottle

Question

How can you make a simple compass?

Setting the Stage

Discuss with students the concept of the Earth as a giant magnet.

Materials Needed for Each Group

- medium-sized glass jar
- piece of thread
- bar magnet
- needle
- piece of stiff paper
- pencil
- data-capture sheet (page 26), one per student

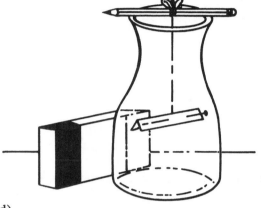

Procedure *(Student Instructions)*

1. Fold the piece of paper into a small tent.
2. Tie the piece of thread around the middle (on the fold).
3. Slide the needle into one side of the paper.
4. Magnetize the needle by stroking it in the same direction about 25 times with a bar magnet.
5. Suspend the needle in the jar by tying the loose end of the thread around a pencil. Place the pencil across the opening of the jar.
6. Observe and record what happens on your data-capture sheet.
7. Bring the magnet close to the needle. Record what happens on your data-capture sheet.

Extensions

- Have students put a piece of cardboard between their magnet and the needle. What happens? Have them keep adding the pieces of cardboard until their magnet no longer affects the needle.
- Have students study early navigation. Have them report their findings to the class.
- Have students find out about magnetic fields of other planets.

Closure

In their magnetism and electricity journals, have students write a short story about a child who is lost on the prairie. There are no landmarks to keep from walking in circles. Have them use what they have learned about making a compass to help the child walk to safety.

The Big Why

The suspended needle aligned itself with Earth's magnetic field. Not all planets or moons have a magnetic field. When it sends satellites to other planets, one of the things NASA is investigating is the magnetic field. Io, a moon of Jupiter, is strongly affected by the large magnetic field of Jupiter.

Compass in a Bottle *(cont.)*

In the space below, draw a picture of the compass that you made. After you placed the needle into the bottle, what did you observe?

Why did this happen?

_____

_____

_____

In which direction did the needle line up?

_____

_____

_____

What happened to the atoms in the needle when you ran a magnet along it?

_____

_____

_____

Which North Is Which?

Question

What is the difference between magnetic north and geographic north (or true north)?

Setting the Stage

Explain to students the difference between true north which is the direction of the North Pole (the point of the axis of rotation), and magnetic north which is 11° from true north.

Materials Needed for Each Group

- compass
- round stick about 2.5" (75 cm) long (A shorter stick will need to be used during the winter months.)
- a marker
- compass and ruler
- large piece of stiff cardboard
- data-capture sheet (page 29), one per student

Note to the teacher: This experience requires a sunny day.

Procedure *(Student Instructions)*

1. Put a piece of cardboard out on the grass where it will be in the sunlight.
2. Use the compass to draw a line indicating magnetic north. Make the line long enough to reach from one edge of the cardboard to 2/3 of the way across.
3. Drive the stick into the ground at the edge of the cardboard at the originating point of your magnetic north line.
4. Leave the stick out there from 11 a.m. until 1 p.m.
5. Mark the top position of the shadow that the stick makes every 15 minutes.
6. After you have made your marks, find the point where the shadow was the shortest. Connect that point with the originating point of the magnetic north line.
7. Compare the two lines and record your observations on your data-capture sheet.

Extensions

- Have students try this on different days. See if they get the same results.
- People tried reaching the North Pole for many years. Have students find out who was first. Ask them what problems early explorers had in their quest for the North Pole.
- There has been some speculation that North America and Europe were once joined at their northern extremes. Have students find out all the information they can about this super continent.

Which North Is Which? *(cont.)*

Closure

On the wall of the classroom, find and mark the spots that are true north and magnetic north. Have students decorate these spots with words that are associated with each. For example, *true north:* North Pole, axis of rotation; *magnetic north:* compass, direction.

The Big Why

You need to measure during the two-hour period because our noon is not always at the sun's highest angle. We change our clocks twice a year! The smallest shadow in this two-hour period always points to true north, because as the Earth rotates about its axis, the sun will be at its highest point when the shadow is the shortest. This alignment always points toward the axis. You can demonstrate this with a globe and a flashlight.

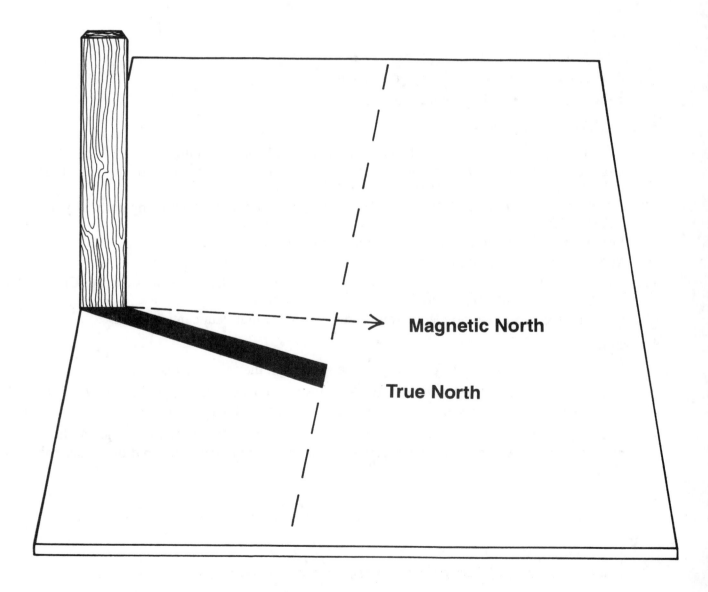

Which North Is Which? *(cont.)*

Answer the questions.

Observations

Measure the angle between true north and magnetic north. _____

How far off were you from the accepted value of 11° ? _____

Conclusions

Why do you think your value was not exact? _____

_____

_____

Would you get the same results if you lived elsewhere? _____

_____

_____

Does the chubby guy in a red suit, who comes around at Christmas, live at true north or magnetic north? If you cannot find this information, give it your best guess and explain why you think this is so. _____

_____

_____

Which method do you think works best for navigating at sea—using a compass (magnetic north) or using the North Star (true north)? Why? _____

_____

_____

_____

_____

_____

Just the Facts

There are two types of electricity, *static* and *current.* When the electrons are in motion, an electric current is created. When they stay attached to an atom, the electricity is said to be static. The modern world depends heavily upon generated electricity to power homes, offices, factories, transportation, and many other things. Electric current also has two types of flow—*alternating* and *direct.* Our homes, factories, and most other major users of electricity use alternating current. Direct current can be produced by batteries.

Following are some facts about electricity:

- Certain materials allow electricity to flow freely. These are called *conductors.* Copper wire is a good example.

- Certain materials impede the flow of electricity. These are called *resistors.* Rubber would be a good example.

- Electricity can be generated chemically. Batteries demonstrate this.

- Lightning is a very powerful form of electricity.

- Electrical energy gives off heat in electric heaters and is also used to light bulbs.

- For electricity to flow, there must be a complete circuit.

- Static electricity is created when two objects are rubbed together.

Battery Power

Question

How can you make a simple battery?

Setting the Stage

- Discuss with students all the ways we use batteries. Bring different kinds and sizes of batteries to show your class.
- Discuss with students what kind of current is present in a battery.

Materials Needed for Each Group

- two wires with alligator clips on the end
- a fat beaker or glass jar
- a zinc electrode 1" x 3" (2.5 cm x 7.5 cm)
- a copper electrode 1" x 3" (2.5 cm x 7.5 cm)
- enough lemon juice to almost fill the jar
- an ammeter or multimeter
- data-capture sheet (page 32), one per student

Procedure *(Student Instructions)*

1. Pour the lemon juice into the jar until it is nearly full.
2. Clip the two wires to the terminals on the ammeter or multimeter.
3. Put the zinc and copper electrodes in the jar as close together as possible without touching.
4. Connect one lead to the copper electrode and the other to the piece of zinc electrode.
5. Record your observations on your data-capture sheet.

Extensions

- Have students repeat this experience, this time using distilled water. What happens? Why?
- Have students repeat this experience, this time adding salt to the distilled water. What happens?
- Have students repeat the experience using orange juice, soda pop, and other liquids.
- Have students find out what a battery back-up system is. Where are these systems found? Why are they important?

Closure

Have your class make a bulletin board that shows the battery made in the experience. Connect it to pictures made by your students of all the different products for which we use batteries.

The Big Why

The lemon juice reacts chemically with the zinc. In the process, it strips the zinc of positively charged ions (atoms with too many or too few electrons; thus they are charged). Now the zinc electrode has an excess of negatively charged electrons. They travel through the wire and end up at the copper electrode. The copper is picking up positive ions from the lemon juice. It becomes the positive electrode.

Some liquids are good conductors of electricity. That is because they have ions in them that are free to move through the liquid. Salt dissolved in water changes to a positive sodium ion and a negative chlorine ion. Salt is an electrolyte; it can conduct electricity. Distilled water will not conduct electricity. Other examples of electrolytes are acids, salts, or alkalis added to water or other solvents.

Battery Power *(cont.)*

In the space below, draw a picture of the battery that you made. Label all the parts.

What did you observe when the battery was connected to the ammeter or multimeter? How did you know that you had made a battery? _____

_____

_____

What does a battery do? _____

_____

_____

What is an electric current?_____

_____

_____

Keeps On Going

Question

Does the Energizer® battery really beat all the others?

Setting the Stage

Discuss with students the Energizer® bunny commercials. Are the claims made by the company really true?

Materials Needed for Each Group

- a variety of different manufacturers' D-cell batteries
- insulated wire about 1' (30 cm) long with the insulation stripped from both ends
- 3-volt light bulbs, one for each brand of battery
- some tape, (Electrical tape works best.)
- data-capture sheet (page 34), one per student

Procedure *(Student Instructions)*

1. Make sure all the batteries you are testing are the same variety. Make sure they are all alkaline or all simple D-cells.
2. Wrap the end of a piece of wire around the base of the bulb.
3. Tape the other end of the wire to the bottom of the battery.
4. Tape the lights down to the top of each battery. Make sure the bottom end of the light bulb touches the top terminal of the battery. This should be done to each battery at the same time. If the class time comes to an end before the light goes out, open the circuit by removing the bottom wire. Resume the experience the next day.
5. Complete your data-capture sheet.

Extensions

- Have students give the batteries a rating that represents how long they lasted.
- Have students write to the company whose batteries were tested and tell them the results of the experience.
- Have students compare a regular battery and an alkaline battery.
- Have students repeat the tests several times and see if the results are the same.
- Have students build a flashlight on their own. They may want to incorporate other materials, such as foil, cardboard tubes, a bulb holder, etc.

Closure

Now that your students have the results of the battery test, compare it with the commercials about batteries that were discussed in Setting the Stage.

The Big Why

Some factors that could affect the battery test are shelf life, where batteries are stored, quantity, and type. Although there are many new battery technologies, alkaline batteries are presently the longest-lasting types.

Keeps On Going *(cont.)*

Record your results on the table and answer the questions.

Brand	Brightness of Light at Beginning	Time It Took To Drain the Power	Rating
1.			
2.			
3.			
4.			
5.			

Which batteries ended up lasting the longest? _____

_____

_____

Did the brightest light last the longest? _____

_____

_____

What factors could influence the test? _____

_____

_____

Which batteries would you buy? _____

_____

_____

Going with the Flow

Question

What materials will allow electricity to flow through them?

Setting the Stage

Discuss with students the concept of a closed circuit. For an electric current to flow, the circuit needs to be complete (or closed). If there is a switch in the circuit or material that does not permit the flow of electricity, the circuit is broken and the flow of electrons is stopped.

Materials Needed for Each Group

- flashlight bulb and holder
- three leads (with alligator clips or insulated wire with stripped ends)
- a variety of different materials (paper clips, pencil, wood ruler, pin, spoon, eraser, etc.)
- data-capture sheet (page 36), one per student

Procedure *(Student Instructions)*

1. Hook the alligator clips or wire leads to each end of the battery, one lead per end.
2. Hook one wire that comes from the battery to one pole of the light bulb holder.
3. Join the third wire to the other pole of the bulb holder.
4. Touch the two unattached ends together. The bulb should light up. Disconnect these two ends. They will be used for testing the materials.
5. Test different materials by connecting them to the two loose terminals.
6. Draw a picture of your circuit, complete the chart, and answer the questions on your data-capture sheet.

Extensions

- Using a 9 volt battery have students repeat the experience. Be prepared to replace the bulb.
- Have students study the safety precautions one should take when using electricity.
- Have the class make a safety poster.

Closure

Have students add their completed data-capture sheets to their magnetism and electricity journals.

The Big Why

Some objects allow the flow of electrons because their atomic design does not hold on to electrons very firmly. This permits them the freedom to travel through the material. These materials are called *conductors*. The atomic bond of other materials is so strong that there are no free electrons; thus, they do not allow the flow of electricity. These materials are called *resistors*.

Going with the Flow *(cont.)*

In the space below, draw a picture of the circuit that you used in the experience.

Record the results of your experience.

	Material Tested	Conducted Electricity	Resisted the Flow of Electricity
1.			
2.			
3.			
4.			
5.			
6.			
7.			
8.			

Why does a circuit need to be complete (closed)? _____

_____

What were the materials made of that conducted electricity? _____

_____

What were the materials made of that resisted electricity? _____

_____

Make a general statement about resistors and conductors and the flow of electricity.

_____

Not-So-Tall Circuit

Question

What is a short circuit?

Setting the Stage

Have students brainstorm about what a short circuit is and what causes it. (The connection does not make a complete circuit for the electricity to flow. When you are dealing with the 120-volt AC current we have coming out of the wall, a spark will often result from a short circuit. This often is the cause of an electrical fire. This will not happen in this experiment.)

Materials Needed for Each Individual

- a battery, D-cell
- battery clips
- flashlight bulbs and holders
- two pieces of insulated wire, about 8" (20 cm) long with the ends stripped
- data-capture sheet (page 38)

Procedure *(Teacher Instructions)*

1. To make a simple battery holder, you need a wide rubber band and some brass brads.
2. Push a brad through the rubber band so that the round end of the brad is on the inside. This will make contact with the battery.
3. Insert another brass brad through the rubber band at the opposite end. Bend the skinny ends of the brads back.
4. When you connect the exposed end of your wire to the brad, wind it around the inside of the rubber band between the round end of the brad and the rubber band.

Procedure *(Student Instructions)*

1. Make the circuits that are illustrated on your data-capture sheet.
2. After you have completed each circuit, record your observations on the line beneath each circuit.
3. Make some of your own circuits.
4. Draw a diagram of them and record your observations on your data-capture sheet.

Extensions

- Have students look up the correct symbols for writing circuit diagrams. Have them draw the diagrams from the data-capture sheet, using symbols.
- Have students read about the early discoveries in electricity and the development of electric power as we use it today.

Closure

In their magnetism and electricity journals, have students write a short story about the day the electricity went off. Where were you and what were you doing when it went off? What did you do to occupy your time when it was off?

The Big Why

To make a complete circuit, the flow of electricity must be unbroken. In some of the circuits used in the experiment, the electricity could not go through the lamp filament; therefore, a closed circuit was not established and the lamp would not light.

Not-So-Tall Circuit *(cont.)*

Make each circuit shown in the pictures. On the line beneath each picture record your observations.

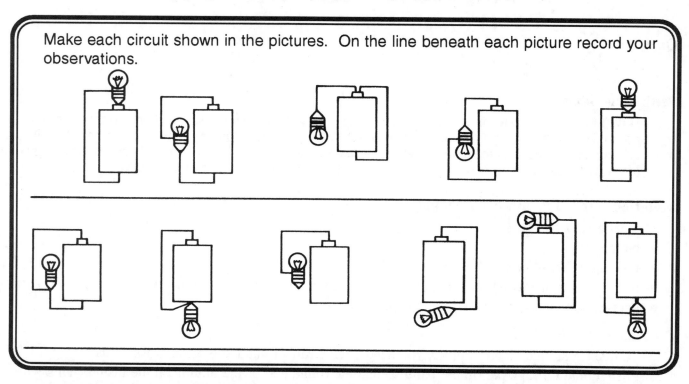

In the space below, draw a picture of a circuit that you made.

Did the circuit make the light come on? If not, how could you change it to make the light come on?_____

_____

_____

_____

From your observations, what is necessary for the light to turn on? _____

_____

_____

38   ©1994 Teacher Created Materials, Inc.

Christmas Tree Puzzle

Question

What is the difference between a *parallel* circuit and a *series* circuit?

Setting the Stage

- Bring to class two sets of Christmas tree lights. Try to find one that is wired in series. (If one bulb goes out, they all go out.) The older sets are wired this way; you can often find them in thrift stores. One set should be wired in parallel. (If one goes out, the rest stay lit.)
- Show your class what happens when one light bulb is taken out of each set. Explain to students this is because of the different way the lights are wired. One method breaks the path so the circuit is not complete.

Materials Needed for Each Group

- a 6-volt battery
- some insulated wire, about 6 to 8 pieces each 6" (15 cm) long with the ends stripped (Alligator clips on the ends will make this lab easier to perform.)
- three light bulbs (3-volt) in holders
- data-capture sheet (page 41), one per student

Procedure *(Student Instructions)*

1. Wire the bulbs to make a series circuit as shown on your data-capture sheet.
2. Observe how brightly the lights are lit. Record your observations on your data-capture sheet.
3. Take one bulb out of its holder. What happens?
4. Wire the bulbs to make a parallel circuit as shown on your data-capture sheet.
5. Observe how brightly the lights are lit. Record your observations on your data-capture sheet.
6. Take one bulb out of its holder. What happens?

Extensions

- If the class has been taught how to draw circuits using electrical symbols, have them draw a parallel and a series circuit with symbols.
- Houses are wired with parallel circuits. Have an electrician come to the class and explain the advantages of this model of wiring.
- Have students see if their parents or an adult will take them to a new home construction site to see how houses are wired.

Christmas Tree Puzzle *(cont.)*

Closure

In their magnetism and electricity journals, have students write a paragraph explaining the two ways to wire a circuit, in series and in parallel.

The Big Why

- The two circuits can be explained through the analogy of two methods of climbing the side of a building. If you use a single rope to get to the top and the rope breaks, you are in big trouble. This is like a series circuit. If you use a ladder to get to the top and one step is broken, you can still make it to the top. This is like a parallel circuit. For household use, the parallel circuit is better. The type of circuit used depends on what work the circuit will be doing.

- The light bulbs in the series circuit will not burn very brightly because they have to share the same power (voltage), so each gets just a small amount. In a parallel circuit each light bulb is connected to the battery itself, so the voltage is not shared; each gets a full amount of power.

Christmas Tree Puzzle *(cont.)*

The first circuit you are to make is a series circuit. Put it together like this diagram:

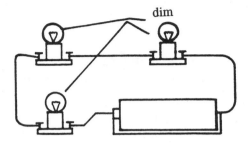

Sketch A: Series Circuit

What did you observe about the glow of the light bulbs? _____

_____

What happened when you removed one light bulb? _____

_____

The second circuit is a parallel circuit. Put it together like the diagram below.

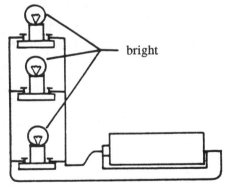

Sketch B: Parallel Circuit

What did you observe about the glow of the light bulbs? _____

_____

What happened when you removed one light bulb? _____

_____

Which circuit is easier to make? _____

_____

Switch It

Question

How does a switch control an electric circuit?

Setting the Stage

Walk over to the wall and turn off the lights; then turn them on again. Ask students how often we use a switch and think nothing of it. How could we turn off the lights without a switch?

Materials Needed for Each Group

- flashlight bulb and holder
- a D-cell battery
- metal strip
- insulated wire with alligator clips on the ends or stripped ends
- a small block of wood
- two thumb tacks or wood screws
- data-capture sheet (page 43), one per student

Procedure *(Student Instructions)*

1. Put a tack into one end of the small wood block.
2. Run a wire underneath the tack, connect its other end to one end of the light bulb holder.
3. Run another wire from the other end of the light bulb holder to the battery.
4. Push the second tack through your small metal strip and part way into the wooden block. It should be close enough to the first tack that the strip of metal can be made to touch both tacks. Run a wire from the other battery terminal around the second thumb tack.
5. Make the metal strip touch the two thumb tacks. Observe and record what happens on your data-capture sheet.
6. Move the metal strip so that it touches just one of the thumb tacks. Observe and record what happens on your data-capture sheet.

Extensions

- Have students design a switch of their own and make it.
- Have students draw the circuit using electrical symbols.
- Have students research other kinds of switches.

Closure

Bring some switches to class that can be opened up for the class to see.

The Big Why

Switches are a convenient way to open and close a circuit. The switches used in thermostats are bulbs of mercury. The mercury conducts electricity and is used to complete the circuit. When a coil of metal gets hot and expands, the bulb will move so that the mercury breaks the electric connection. If you have an old thermostat, bring it to class to show the children this kind of switch.

Switch It *(cont.)*

In the space below, draw a picture of the circuit that you made. Use any electric symbols that you know.

What is the purpose of a switch in an electrical circuit? _____

_____

Think of all the things that you do in a day. List the times that you turn on or off a switch.

Don't Con Fuse Me

Question

How do fuses work?

Setting the Stage

- Bring in some examples of fuses and circuit breakers. Explain that a fuse burns out, but a circuit breaker is designed to disconnect the circuit.
- Discuss with students the reason that we use fuses and circuit breakers in a circuit.

Materials Needed for Each Group

- a 9-volt battery
- a board
- insulated wire with the ends stripped
- steel wool
- two thin 1" (2.5 cm) nails
- a hammer
- data-capture sheet (page 45), one per student

Note to the teacher: The purpose of this experience is to melt the steel wool so that the circuit is disconnected. The steel wool will get very hot. Warn students to be careful.

Procedure *(Student Instructions)*

1. Hammer the nails into the board about 1" (2.5 cm) apart. Leave about half of the nail exposed.
2. Wire a strand of steel wool between the two nails.
3. Connect one of the wires to one of the battery terminals; join the other end to one of the nails.
4. Connect a second wire to the second battery terminal.
5. Touch this to the second nail. Be careful — the steel wool is HOT!
6. Observe what happens and record on your data-capture sheet.

Extensions

- Have students repeat the experience, this time with a thicker strand of copper wire between the nails. What happens?
- Ask the city engineering department what the laws are concerning the use of fuses in home wiring. Then share those laws with your students.

Closure

In their magnetism and electricity journals, have students write a paragraph explaining how fuses work and why they are used in electric circuits.

The Big Why

Fuses are safety systems that are used when too much current is flowing through a system. Too much current can destroy electrical equipment such as radios or computers. Most modern houses have circuit breakers which serve as fuses. A circuit breaker is like a switch that is turned off when too much current is going through the circuit. The advantage of a circuit breaker is that you do not need to keep a supply of new fuses on hand. Many computer users have surge protectors between the electrical outlet and their computers. This gives them added protection against current surges. A surge does not always burn out a fuse, but it can harm the delicate circuitry of a computer.

Don't Con Fuse Me *(cont.)*

In the space below draw a picture of the circuit that you used.

What did the steel wool represent in the circuit? _____

_____

What happened to the steel wool when the circuit was completed? What other materials could we use to make a fuse? _____

_____

Why do you think the wiring of a fuse is encased in glass or special plastic? _____

_____

Where would fuses be useful? _____

_____

In this experience electricity was turned into heat. What appliances make use of this concept to our benefit? _____

_____

Spark in a Cup

Question

What is a *capacitor?*

Setting the Stage

Discuss with students the concept of storing electricity. Batteries can be considered as one method of storing electricity. Have your class brainstorm other ways they might store electricity. What would be the advantage of doing this?

Materials Needed for Each Group

- a plastic cup
- a plastic ruler
- a piece of wool
- two pieces of copper wire, about 4" (10 cm) longer than the cup and container are high
- a plastic container that is larger than the cup
- a plastic clothespin
- data-capture sheet (page 47), one per student

Procedure *(Student Instructions)*

1. Fill the plastic cup three-fourths full of water and place it in the plastic container.
2. Put water into the container until it is at a level equal to the water in the cup.
3. Make a small stand at the bottom of the two pieces of copper by bending one end into a circle. They need to be able to stand in the water. Place one of the pieces of copper in the cup and the other in the container. The wire should be above the water level.
4. Charge the copper wire that is in the cup. To do this, rub the ruler vigorously with the wool cloth. Touch it to the piece of copper wire in the cup.
5. Repeat this step about 15 times.
6. Now take the clothespin and push the outer wire towards the inner wire.
7. Observe and record what happens on your data-capture sheet.
8. Repeat this experience several times.

Extensions

- Have students try other liquids (distilled water, salt water, soda pop, orange juice, etc.). Does the liquid affect the ability to create a spark?
- Have students research the use of capacitors in an electric circuit. Have them report their findings to the class.

Closure

Have students add their completed data-capture sheets to their magnetism and electricity journals.

The Big Why

The water in each section collects charges. These sections are separated by the plastic cup. The charged ruler imparts its electrons to the wire in the center section. Therefore, it is negatively charged. When the outside wire is brought closer to the inner wire, a spark jumps across. The electrons transferred to the neutral wire. This is similar to a spark jumping from one person to another when you walk barefooted across a rug on a dry day and then touch another person.

Spark in a Cup *(cont.)*

In the space below, draw a picture of the capacitor that you made.

What did you observe? _____

_____

What caused the spark to jump across? _____

_____

Where did the electricity collect? _____

_____

Do you think changing the liquid would make a difference? _____

_____

What needs to be done to collect a greater charge? _____

_____

Make a rule for what you observed. Share it with the class.

_____

_____

Just the Facts

Hans Christian Oersted was doing an experiment with electricity when he accidentally discovered a link between electricity and magnetism. When news of this experiment got out, many other scientists began to experiment with this new phenomena. However, it was the English scientist Michael Faraday who made the most important discovery in electromagnetism. While doing some experiments, Faraday discovered that the movement of magnetic lines of force across a wire creates a current. Faraday was able to induce a current in a wire. His discovery of electromagnetic induction has had a great impact on our world. The principle of electromagnetic induction allowed for the building of huge electrical generators that power our modern society.

Following are some facts about electromagnets:

- Light is a form of an electromagnetic wave.

- Electricity can be used to make powerful magnets called *electromagnets*.

- Electromagnetic induction allows for the creation of huge electrical generators.

- Electromagnets can be switched on and off.

- The strength of an electromagnet may be increased by adding more electrical current to it.

- Without electromagnetic devices, there would be no hair dryers, televisions, stereos, or computers.

A Mysterious Force

Question

Does an electric current produce a field?

Setting the Stage

Research the life of Hans Christian Oersted. Share some of his life with your class. Sometimes great discoveries are found by accident. Hans Christian Oersted (1820's) was experimenting with a battery and a piece of wire. A compass was lying nearby. What he discovered by chance is what we are going to see today.

Materials Needed for Each Group

- a D-cell battery
- copper wire with the ends stripped, about 12" (30 cm) long
- a magnetic compass
- masking tape or electrical tape
- data-capture sheet (page 50), one per student

Procedure *(Student Instructions)*

1. Connect one end of the wire to one pole of the battery.
2. Form the wire into a large loop. Place the compass inside the loop, close to the wire.
3. Touch the loose end of the wire to the other pole of the battery. Observe and record the results on your data-capture sheet.
4. Place the compass outside of the loop, close to the wire.
5. Touch the loose end of the wire to the pole of the battery again. Observe and record the results on your data-capture sheet.
6. Slowly move the compass away from the wire. Measure how far you need to move it before the field produced by the wire has no effect on the compass.
7. Complete your data-capture sheet.

Extensions

- Have students study the life of Hans Christian Oersted.
- Have students research the effects of electromagnetic fields on the human body.

Closure

Have students add their completed data-capture sheets to their magnetism and electricity journals.

The Big Why

One of the characteristics of moving electrons is that they generate a magnetic field. This magnetic field has no north and south poles; it is a continuous field. The field lines wrap around the wire like rings. The left-hand rule is used to determine the direction of the field. Place your left hand so that your thumb is pointed in the direction that the current flows (from a negative pole to a positive pole). Curl your fingers and you will have the direction of the field lines. When the wire is placed on a table, the field lines seem to go into the table on one side of the wire and come out of the table on the other side of the wire. This is why the compass will point in different directions when placed on different sides of the wire. The field points up on one side and down on the other side. This is an electromagnetic field.

A Mysterious Force *(cont.)*

In the space below, draw a picture of your circuit. Draw the compass in the two spots that you placed it. Show the direction that the compass pointed in each case.

What is the reason the compass needle moved? _____

_____

What would the compass needle do if a magnet was brought close to it? _____

_____

What did the compass needle show about the area around the wire? What kind of field is produced by the wire? _____

_____

From this experiment make a general rule about wires with current flowing in them and the field that is created. _____

_____

Force Has It

Question

What is a galvanometer and how does it work?

Setting the Stage

- Review with students the results of the experience "A Mysterious Force" (pages 49-50). We now know that current and magnetism are connected; can we use one to measure the other?
- Let your class suggest some possible ways this might be accomplished.

Materials Needed for Each Group

- a small compass
- bell wire
- two alligator clips
- cardboard
- plastic ruler
- masking tape
- medium-sized can
- several different sizes of batteries
- data-capture sheet (page 53), one per student

Procedure *(Student Instructions)*

1. From the cardboard cut out a circle about 4" (10 cm) in diameter. Make two notches on opposite sides of the cardboard.
2. Tape the circle onto the end of a ruler.
3. Wind the wire 10 to 12 times around the can to form a coil. Tape the loops of the coil together in two places. Leave two long ends of about 12" (30 cm).
4. Place the coil over the circle and into the notches so that it is perpendicular with the cardboard. (See drawing on your data-capture sheet.)
5. Tape a small compass to the center of the cardboard. Line up the needle with the coils.
6. Tape the ruler to the edge of a table; be sure the needle is still lined up with the coils.
7. Attach the free ends of the wire to alligator clips.
8. Attach the alligator clips to the poles of a battery.
9. Observe and record the reaction of the compass needle on your data-capture sheet.
10. Trade locations of the alligator clips. Record what happens on your data-capture sheet.

Extensions

- Have students learn how to use a small meter to test voltage and current.
- Have students research the life of Luigi Galvani (1790) and report their findings to the class.
- Have students make a list of different kinds of meters and their uses.

Force Has It *(cont.)*

Closure

Have students add their completed data-capture sheets to their magnetism and electricity journals.

The Big Why

- When an electric current travels through a wire, a magnetic field is produced. To increase the field we made the wire into a coil; the more coils, the stronger the magnetic effect. This results in a more sensitive meter. The field deflects the compass needle when a battery is connected to the coil. The more current the battery produces, the larger the magnetic field and the more the needle will deflect.

- The class will notice that sometimes the needle is deflected in one direction and sometimes in the opposite direction. This is caused by the direction of current flow. If you exchange the alligator clips on the battery, the needle moves in the opposite direction.

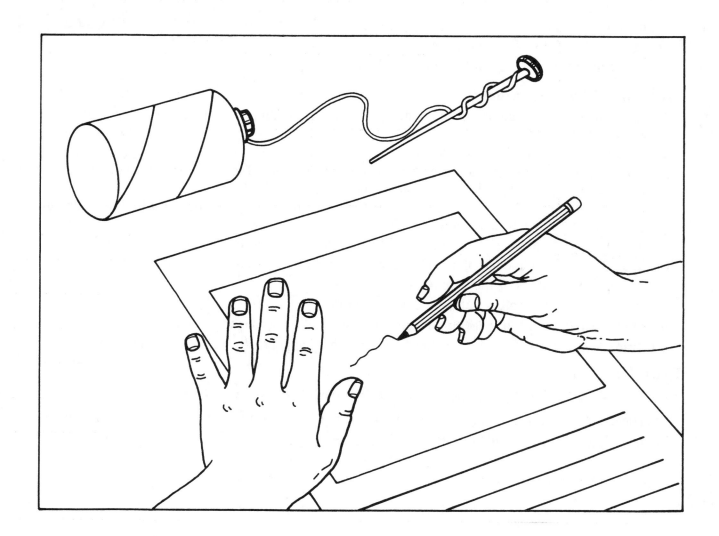

Force Has It *(cont.)*

Build your galvanometer to resemble the picture below.

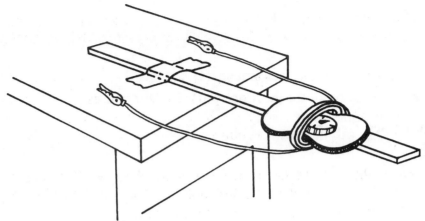

Record your observations in the table.

	Battery Size	**How Far the Needle Moved**
1.		
2.		
3.		
4.		

What causes a compass needle to move under normal conditions? When you connect the battery, what does the coil become? _____

_____

What is your galvanometer measuring? _____

_____

Which battery produced the greatest deflection of the compass needle? _____

_____

What happened when you changed the places of the alligator clips? Make a general statement or rule about what you have learned in this experience. _____

_____

Great Nail Attractor

Question

How can we test the strength of an electromagnet?

Setting the Stage

- Tell students some of the uses of electromagnets.
- Have students make a list of all the electromagnets in their houses.

Materials Needed for Each Group

- a 6-volt battery
- insulated 20-gauge copper wire, about 1 yard (1 m) with ends stripped
- paper clips
- two thumbtacks
- a small piece of wire
- a small wooden board
- a ruler
- data-capture sheet (page 56), one per student

Note to the teacher: Save the switch for use in Ding Dong Bell (pages 64-66).

Procedure *(Student Instructions)*

1. Coil the yard (m) of wire around the nail, leaving about 8" (20 cm) of wire free at each end.
2. Put a tack at each end of the small block of wood.
3. Attach one end of the wire coming off the nail to one of the tacks. Open up one paper clip and hook one end under the same thumbtack.
4. Attach the small wire to the other thumbtack and to a battery terminal.
5. Hook the other long wire coming off the other end of the nail to the other battery terminal.
6. Lay the ruler down on the table and place a paper clip at the 11" (27.5 cm) mark.
7. The paper clip that is part of the circuits acts as a switch. Complete the circuit by closing the switch — touch the paper clip to the other tack.
8. Place the nail at the zero end of the ruler. Slowly move the nail towards the paper clip. Record the distance from the paper clip when the magnetic nail begins to attract the clip.
9. Repeat the above step five times.
10. Find the maximum number of paper clips that the nail will pick up. Turn the switch off. What happens? Repeat five times.
11. Complete your data-capture sheet.

Extensions

- Have students repeat the experience, this time wrapping the nail with fewer loops.
- Have students repeat the experience, this time wrapping the nail with more loops.
- Have students repeat the experience, this time adding another battery to their circuit. Have them make sure the poles are + to + and - to -.
- Have students try different cores to wrap the wire around. Which one produces the strongest magnetic attraction?

Great Nail Attractor *(cont.)*

Closure

Discuss with students the advantages of having a magnet that you can turn on and off. What would be some uses of this kind of magnet?

The Big Why

When a current flows through a wire, a magnetic field is generated. By coiling the wire around the nail, the electromagnetic field lines up the poles of the iron in the nail and it becomes a temporary magnet. The two enhance each other to form a more powerful magnetic attraction. More coils will increase the strength of the field, as will an increase in current.

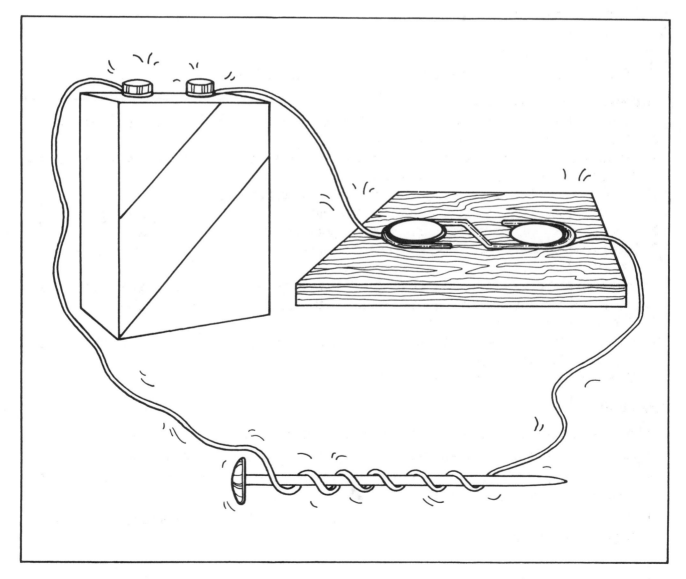

Great Nail Attractor *(cont.)*

In the space below draw a picture of the circuit you used.

Trials	Number of Inches (cm) from the paper clip	Number of paper clips that can be lifted
Trial 1		
Trial 2		
Trial 3		
Trial 4		
Trial 5		
Average		

Conclusions

What made the coil become a magnet? _____

What would make the magnet stronger? _____

Write a sentence about what you have learned concerning electromagnets. _____

A Magnetic Picker-Upper

Question

How can we build an electromagnetic crane?

Setting the Stage

Review with students the activity "Great Nail Attractor" (pages 54-56).

Materials Needed for Each Group

- a 6-volt battery
- a shoe box
- a thin, long cardboard box
- two empty spools from thread
- two pencils
- a piece of strong thread
- a long insulated wire with alligator clips on both ends
- a second long wire with ends stripped
- a 2-3" (5-7.5 cm) iron nail
- data-capture sheet (page 60), one per student

Procedure *(Student Instructions)*

1. Refer to the diagram as needed (page 59).
2. Stand the shoe box vertically upright.
3. Place the battery in the bottom of the box.
4. Cut off the ends of the long thin box.
5. Glue the thin box to the shoe box so that its end sticks over the shoe box end by 3" (7.5 cm).
6. Put a spool inside the long thin box at the back end. Run a pencil through the side of the box, through the spool, and out the other side of the box. Turn the pencil a few times to make sure it turns freely.
7. Put the other spool inside at the front end of the long thin box. Run a pencil through the side of the box, through the spool, and out the other end of the box.
8. Tie off a piece of thread around the rear spool, tape it in place; and let the rest of the thread hang down over the top of the front spool.
9. Tie the loose end of the thread around the head of the nail so that it hangs down and barely touches the ground.
10. Run a wire from one lead of the battery to the nail and coil it around the nail about twenty turns. Make sure the stripped end of the wire is in contact with the nail and near the tip of the nail.
11. Clip one end of the alligator clip to the point of the nail where the first wire ends. Run the other end of the wire through the long thin box and down to the battery.
12. When you are ready to operate the crane, hook the unused alligator clip to the open terminal of the battery.
13. Lower your crane into a pile of paper clips or other objects.
14. Raise the crane by turning the pencil at the rear spool.
15. Answer the questions on your data-capture sheet.

A Magnetic Picker-Upper *(cont.)*

Extensions

- Have students find ways to make their magnet more powerful. Have them use their knowledge of electromagnets to do this.
- Have students research where electromagnetic cranes are used in industry. Have them share their findings with the class.

Closure

Share your creations with other classes or put them on display in the library with a poster explaining about electromagnets.

The Big Why

The current running through the wire and around the nail creates an electromagnet. This is an example of an electromagnet put to use.

A Magnetic Picker-Upper *(cont.)*

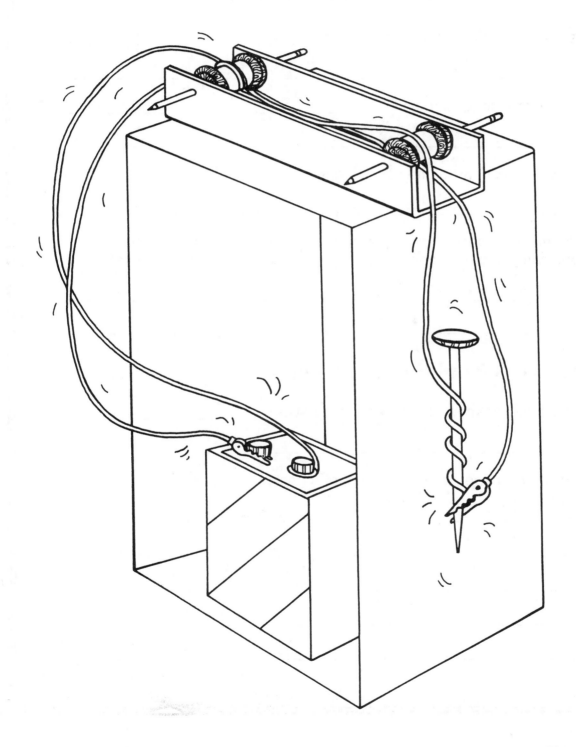

A Magnetic Picker-Upper *(cont.)*

Answer the questions.

A Construction Diary

What was the hardest part of the construction? _____

_____

Which part took the longest to do? _____

_____

What was the easiest part of the construction? _____

_____

If you were making a second crane, what would you do differently? What suggestions can you give to another student who plans to make the crane?

_____

_____

The Science Behind the Construction

List the facts that you know about electromagnets. _____

_____

On the back of this paper, make a sample of a poster that explains electromagnets.

_____

_____

_____

_____

_____

Electromagnetism and Plant Growth

Question

Does electromagnetism have any effect on the growth of lima bean seeds?

Setting the Stage

Discuss with students things that affect the growth of plants; include water, sun, soil conditions, weather, pollution, and music.

Materials Needed for Entire Class

- seven plastic cups
- a roll of insulated copper wire
- 6 six-volt batteries to start (More will be needed; rechargeable batteries are recommended.)
- lima bean seeds
- paper towels
- data-capture sheet (page 63), one per student

Procedure *(Teacher Instructions)*

1. There will be one classroom set-up for this experience. However, the experience contains six identical parts, and one plant which will be grown as a control with no electromagnetic field.
2. Divide the class into six groups and let each group put together part of the experience.
3. The wire will need to be stripped at the ends after each group has coiled it around the cups.
4. The batteries will have to be changed at least once in the morning and once in the afternoon as late as possible.
5. Continue the experience as long as you think necessary. The sprouts should be at least 1" (2.5 cm) long before you quit.

Procedure *(Student Instructions)*

1. Coil the wire around the cup. Group #1: three coils. The remainder of the groups will wrap ten coils, 15 coils, 20 coils, 40 coils, and 100 coils respectively.
2. Line the inside of the cup with a wet paper towel. Put three lima beans between the paper towel and the outside of the cup.
3. Connect the loose ends of the wire to a battery.
4. Prepare one cup with wet paper towel, seeds and no coils as a control.
5. Place near a window.
6. Observe the plants once a day.
7. Make sure the paper towels stay moist.
8. Record your observations on your data-capture sheet.

Extensions

- You may want to have students try to grow different things under these conditions.
- Have students research other experiments that have been done with plant growth. Examples: growing plants without soil, and learning how plants behave in weightless conditions of space.
- Have students find news articles about the effects on the human body of electromagnetic fields from high tension lines.

Electromagnetism and Plant Growth *(cont.)*

Closure

Have students add their completed data-capture sheets to their magnetism and electricity journals.

The Big Why

Living material is composed of tiny molecules, many of which are polarized—that is, one side of the molecule is more positively charged than the other, causing the molecule to behave like a miniature magnet. Some chemical reactions depend on this characteristic. The coil will create a magnetic field in which the plant will grow. Depending on the strength of the field and the type of plant, you should see a difference in the plant growth.

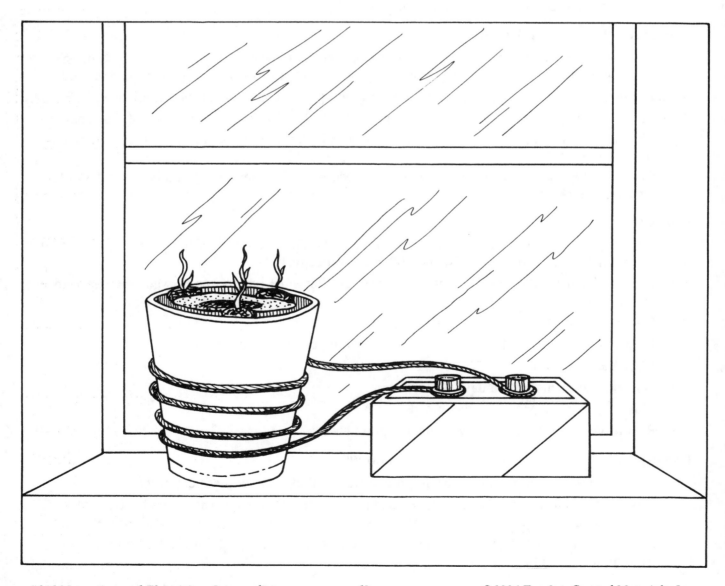

Electromagnetism and Plant Growth

(cont.)

Observe the plants each day. Record your observations below. Begin recording on the day that you first notice any change in the seeds.

Day of Observation	Control	Cup Number	Observation Conclusions

Which plant grew the fastest? _____

_____

Which plant looked the healthiest? _____

_____

Why is it important to have one plant growing without any electromagnetic field as a control?

_____

What did your observations teach you about plants and electromagnetic fields? _____

_____

Ding Dong Bell

Question

How can you turn electricity into sound?

Setting the Stage

Discuss with students what causes sound waves, how sound waves travel, and how the ear responds to sound waves.

Materials Needed for Each Group

- switch from "Great Nail Attractor" (pages 54-56)
- 6-volt battery
- three pieces of wire
- simple doorbell
- screwdriver
- data-capture sheet (page 66), one per student

  Note to the teacher: If you do not have the switch from "The Great Nail Attractor" (pages 54-56), you may make one using the following:

- a small piece of wood
- two thumbtacks
- a paper clip

Procedure *(Teacher Instructions)*

1. Place a thumbtack into the wood.
2. Open up the paper clip and hook one end under the thumbtack.
3. Place the second thumbtack close enough to the first so that the paper clip can touch it.
4. The switch is on when the clip touches both tacks. To turn the switch off, rotate it so that the contact is broken.

Procedure *(Student Instructions)*

1. Remove the doorbell cover and expose its inner workings.
2. Hook a wire to the battery and run it to a contact point on the bell.
3. Run another wire from the battery to a terminal on the switch.
4. Connect the other terminal on the switch to the contact point on the bell.
5. Press the switch.
6. Complete your data-capture sheet.

Extensions

- Have students investigate other ways that electricity is changed and what it changes into.
- Have students research the invention of radios. Have them report what they learned to the class.
- Have students make a simple radio from a kit.

Closure

Have your class make a bulletin board that shows the diagram of the experimental circuit, shows the sound waves from the bell as they travel through the air, and a cutaway view of the human ear with the major parts labeled.

Ding Dong Bell *(cont.)*

The Big Why

Closing the switch forms a complete circuit. The coil turns into an electromagnet and jerks the striker on the bell. When the striker hits the bell, it breaks the circuit. The striker flips back again and makes a closed circuit. This happens many times in a second. The striker tapping the bell sounds like a ring.

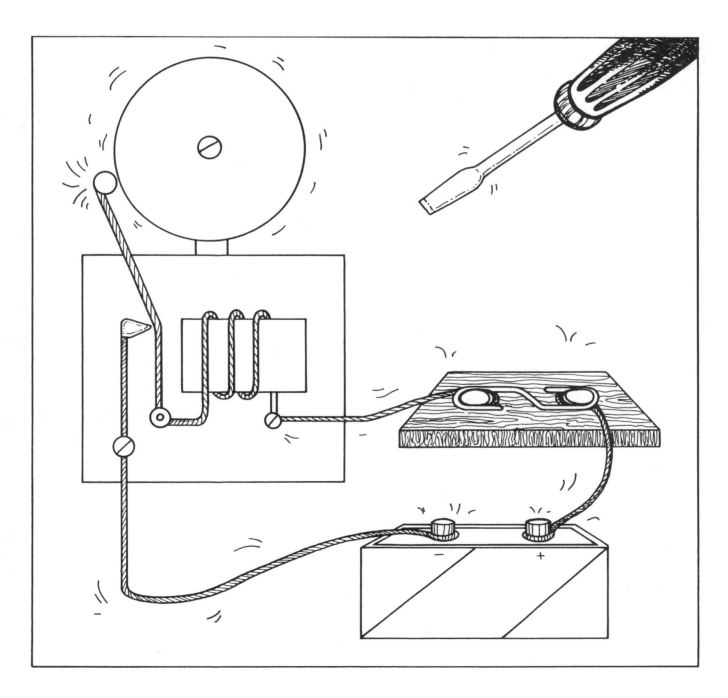

Ding Dong Bell *(cont.)*

In the space below, draw a picture of the bell circuit.

Explain how the electricity was turned into sound waves. _____

_____

How did the sound waves travel to your ear? _____

_____

List different kinds of bells (or ways that bells are used). _____

_____

_____

Motor Time

Question

How can you build a very simple electric motor?

Setting the Stage

- Have a class discussion about motors.
- Have students make a list of all the things they use that have a motor.

Materials Needed for Each Group

- two square magnets
- copper wire
- 6 volt battery
- tape, masking or electrical
- paper clips
- a small block of wood
- data-capture sheet (page 68), one per student

Procedure *(Student Instructions)*

1. Wind the copper wire around any small cylinder so that you get a coil. Leave about 4" (10 cm) of wire at each end. Wrap a small piece of tape at two places on the coil so it does not unwind.
2. Tape the two magnets to the board.
3. Attach the paper clips to the battery as shown in the illustration.
4. Put the coil ends in the paper clip holders.
5. Start the motor by giving the coil a spin.

Extensions

- Have students try to make a larger coil. Ask them what happens when there is a larger coil.
- Have them wire another battery into the circuit. What happens?
- Have students research how an electric train runs. Share that information with the class.
- Have students research electric vehicles. What are their motors like?

Closure

Have your class make a display explaining how motors are used. Place it and your motors in a central area so that they can be shared with other classes.

The Big Why

The wire coil has an electric current running through it which creates a magnetic field around it. The magnet alternately repels and attracts the coil, causing it to rotate. Motors change electricity into motion.

Motor Time *(cont.)*

Draw a picture of the motor that you made. Label the parts.

Why does the coil turn? _____

_____

List the different ways that motors are used. _____

_____

_____

_____

Make Mine a Turbine

Questions

- Where does the electricity that we use come from?
- What does a turbine have to do with the answer?

Setting the Stage

Discuss with students power plants and how they operate. See the drawing below and The Big Why for information.

Materials Needed for Each Group

- piece of cardboard
- turbine pattern
- tape
- two thumbtacks
- scissors
- pencil
- a long thin nail
- two paper clips
- a wood block
- copy of turbine pattern (page 71)

Procedure *(Student Instructions)*

1. Cut out the turbine pattern. Trace it on the cardboard. Cut out the cardboard turbine.
2. Put a long thin nail through its center.
3. Secure it with tape and slightly twist each blade inward.
4. Bend two paper clips and tack them into the wooden block as shown in the diagram.
5. Place the ends of the nails into the paper clip cradles.
6. Blow on your turbine.

Extensions

- Take your class on a field trip to a power station.
- Have someone from the electric company give a presentation to the class.
- Do a class study of Thomas Edison and the power station that he built.
- Have students present reports on power sources and the environmental problems that each kind might cause.

Closure

Make up a play (The Power Play) in which each group is a part of a power station (Steam Boiler, Turbines, Electrical Generator, Steam Condenser, Transformers, High Tension Lines, Substations). Have each group write dialog to explain what happens at each location.

Make Mine a Turbine *(cont.)*

The Big Why

- The electricity used in homes and industry is generated at man-made power stations. There are three main types of power stations that generate electricity. One type uses fast flowing water to turn the turbines, the second and most common type uses oil or coal to generate steam that runs the turbines, and the third type uses nuclear energy.

- The basic premise is that turbines are turned to power an electrical generator. The electrical generator spins a magnet between coils of wire. The bigger the magnet and coils, the greater the voltage output. IF the magnet spins faster, that will also increase the voltage.

- The power stations generate the electricity at about 11,000 volts. This is not a high enough voltage to deliver the electricity economically. Therefore, the electricity is sent through a transformer and stepped up to 400,000 volts. Then the electricity is carried over the wires. This voltage is far too high for a house or business. When it reaches an area substation it is stepped-down through another transformer. This lower voltage is then carried to homes and businesses.

- The voltage coming into a home is very high. It has enough power to kill a person. Playing with plugs and sockets is very dangerous.

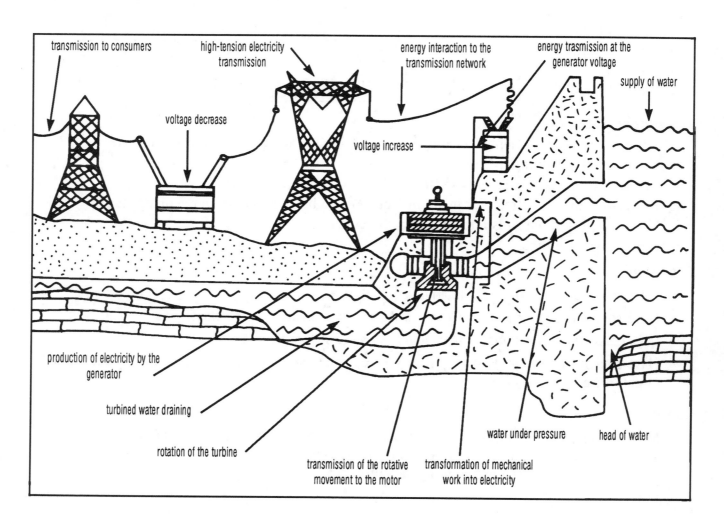

Make Mine a Turbine (cont.)

Cut out the turbine pattern.

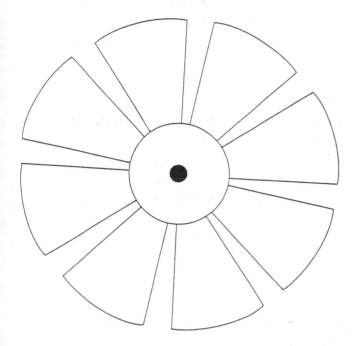
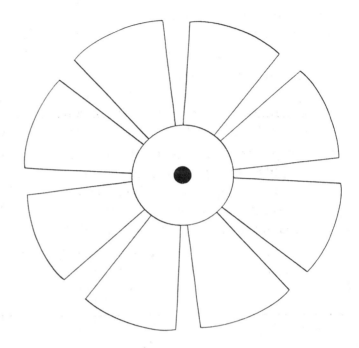
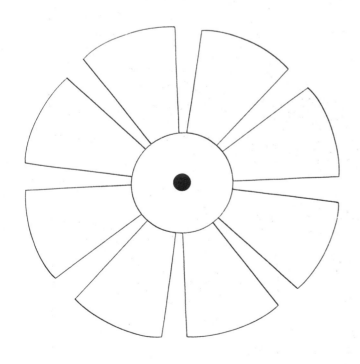
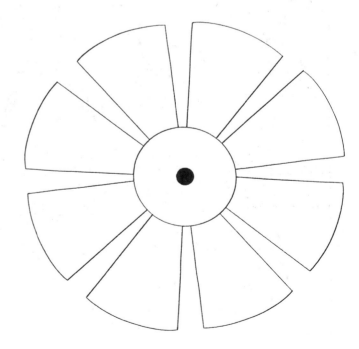

Observe

At this station you will find ten items (aluminum foil, copper wire, eraser, nail, paper, paper clip, pencil, string, cup of water, piece of wood), that will or will not conduct electricity in a simple circuit. Test each item and record the results.

Materials Needed at This Station

- "D" battery
- three pieces of copper wire 6" (15 cm) long
- flashlight bulb and holder
- two alligator clips
- test items (aluminum foil, copper wire, eraser, nail, paper, paper clip, pencil, string, cup of water, piece of wood)
- data-capture sheet (page 73), one per student

Procedure *(Student Instructions)*

1. One at a time attach the alligator clips to each end of the ten items. If the lightbulb lights up, you have created a simple circuit.
2. Observe what happens during each test and record your observations on your data-capture sheet.
3. Put your finished data-capture sheet in the collection pocket on the side of the table at this station.

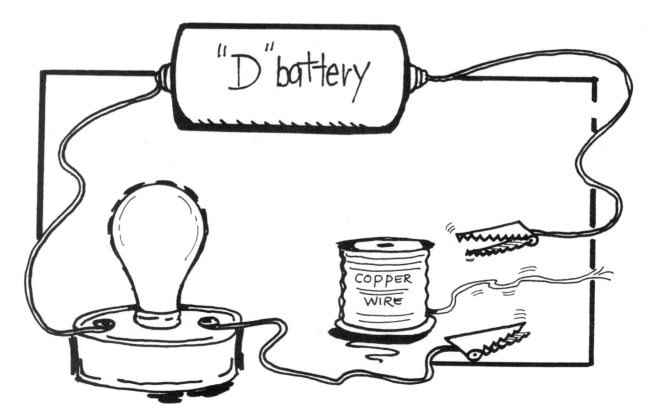

Observe *(cont.)*

Before beginning your investigation, write your group members' names on the lines below.

_____ Project Leader ⠀⠀⠀⠀ _____ Stenographer

_____ Physicist ⠀⠀⠀⠀ _____ Transcriber

Complete the chart.

	Was the circuit complete?	**Yes**	**No**
1	aluminum foil		
2	copper wire		
3	eraser		
4	nail		
5	paper		
6	paper clip		
7	pencil		
8	string		
9	cup of water		
10	piece of wood		

Communicate

Before beginning your investigation, write your group members' names on the lines below.

_____ Project Leader   _____ Stenographer

_____ Physicist   _____ Transcriber

At this station you will find a nail, paper clips, 6-volt battery, and copper wire. If you increase the amount of wire wrapped around a nail, does the magnetic strength increase or decrease? If the wire is wrapped tightly or loosely around a nail, will the magnetic strength increase or decrease?

Procedure *(Student Instructions)*

1. Wrap the copper wire around a nail a random number of turns. Make sure to leave a tail of wire at each end of the nail.
2. Attach the wire ends to each end of a battery.
3. Now try to pick up as many paper clips as possible.
4. Record the information needed on the chart below.
5. Repeat steps 1-4, each time changing the variable (number of wire wraps, tight wrap, loose wrap, etc.).
6. Put your finished activity sheet in the collection pocket on the side at this station.

Number of Wire Wraps	Number of Paper Clips Picked Up

How Wire Was Wrapped	Number of Paper Clips Picked Up

Compare

Before beginning your investigation, write your group members' names on the lines below.

_____ Project Leader   _____ Stenographer

_____ Physicist   _____ Transcriber

At this station you will test to see which part of the magnet is the strongest.

Materials Needed at This Station

- several bar magnets
- spring balance
- large paper clip

Procedure *(Student Instructions)*

1. Place a bar magnet on a table or desk and secure it so it does not move.
2. Attach a paper clip around the hook of a spring balance.
3. To test the pull required to lift the paper clip from the magnet, move the paper clip starting at one end 1/2" (1.25 cm) along the magnet and test the pull at each point.
4. Record this measurement on the chart below.
5. Repeat this with several magnets.
6. Record the reading of your spring scale for each 1/2" (1.25 cm) mark.
7. Put your finished activity sheet in the collection pocket on the side of the table at this station.

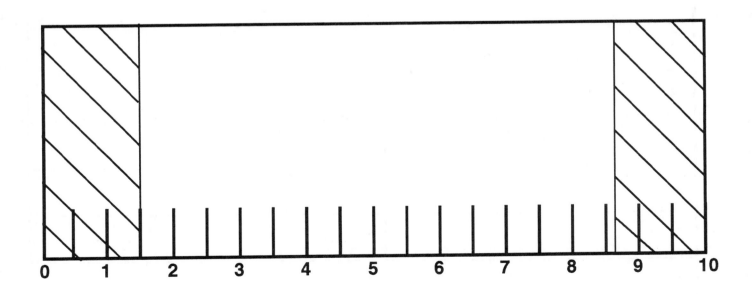

Order

Before beginning your investigation, write your group members' names on the lines below.

_____ Project Leader   _____ Stenographer

_____ Physicist   _____ Transcriber

At this station you will test a variety of batteries to see which one lasts the longest. As the results are noted, record the information on the chart below.

Materials Needed at This Station

- several batteries (same size and voltage)
- flashlight bulb
- wire
- tape

Procedure *(Student Instructions)*

1. Make a simple circuit with each battery, bulb, and wire.
2. Record how many hours each battery lasts. Then record the batteries from best to worst.
3. Put your finished activity sheet in the collection pocket on the side of the table at this station.

	Name of Battery	How Long Did It Last?
1.		
2.		
3.		
4.		
5.		
6.		
7.		
8.		

Categorize

Before beginning your investigation, write your group members' names on the lines below.

_____ Project Leader   _____ Stenographer

_____ Physicist   _____ Transcriber

At this station you will need to decide which objects attract a styrofoam ball and which objects push it away.

Materials Needed at This Station

- styrofoam ball
- balloon
- plastic comb
- ping pong ball
- wood ruler

- thread
- plastic pen
- felt or wool cloth
- plastic bag
- plastic ruler

Procedure *(Student Instructions)*

1. Attach a styrofoam ball to a piece of thread and suspend it from a piece of wood.
2. Rub each item with the wool or felt cloth.
3. After rubbing each item, place it near the styrofoam ball and record on the chart below which items attract the ball and which items repel it.
4. Put your finished activity sheet in the collection pocket on the side of the table at this station.

Item	Attract	Repel

Relate

Before beginning your investigation, write your group members' names on the lines below.

_____ Project Leader _____ Stenographer

_____ Physicist _____ Transcriber

At this station you will find several photographs of different electrical items. Write the items in the columns of listed categories. Some items may appear in more than one place.

Farm	Hospital	Home	Transportation	Factory	Communication

Put your finished activity sheet in the collection pocket on the side of the table at this station.

Infer

Before beginning your investigation, write your group members' names on the lines below.

_____ Project Leader   _____ Stenographer

_____ Physicist   _____ Transcriber

At this station you will find 15 different objects (balloon, drinking glass, eraser, fork, key, nail, nickel, nuts and bolts, paper clip, pencil, penny, piece of plastic, piece of tin, scissors, straight pin, and thumb tack). Using everything you know about magnets, list those that you think will be attracted to a magnet and those that will not be attracted to a magnet.

Attracted to a Magnet	**Not Attracted to a Magnet**
1. _____	1. _____
2. _____	2. _____
3. _____	3. _____
4. _____	4. _____
5. _____	5. _____
6. _____	6. _____
7. _____	7. _____
8. _____	8. _____
9. _____	9. _____
10. _____	10. _____
11. _____	11. _____
12. _____	12. _____
13. _____	13. _____
14. _____	14. _____
15. _____	15. _____

Put your finished activity sheet in the collection pocket on the side of the table at this station.

Apply

Before beginning your investigation, write your group members' names on the lines below.

_____ Project Leader　　　　_____ Stenographer

_____ Physicist　　　　　　_____ Transcriber

At this station you will be creating a continuity tester, using your newfound knowledge of electrical circuits.

Materials Needed at This Station

- several lengths of wire
- flashlight bulb and holder
- battery
- tape
- paper clip
- eraser
- other materials to test (e.g., key, nail, paper, pencil, penny, etc.)

Procedure *(Student Instructions)*

1. Connect the wires, battery, and bulb holder.
2. Leave two live wires so you can test your materials.
3. After testing your materials, place them in the correct column.

Insulators	Conductors

Put your finished activity sheet in the collection pocket on the side of the table at this station.

Science Safety

Discuss the necessity for science safety rules. Reinforce the rules on this page or adapt them to meet the needs of your classroom. You may wish to reproduce the rules for each student or post them in the classroom.

1. Begin science activities only after all directions have been given.

2. Never put anything in your mouth unless it is required by the science experience.

3. Always wear safety goggles when participating in any lab experience.

4. Dispose of waste and recyclables in proper containers.

5. Follow classroom rules of behavior while participating in science experiences.

6. Review your basic class safety rules every time you conduct a science experience.

You can still have fun and be safe at the same time!

Magnetism and Electricity Journal

Magnetism and Electricity Journals are an effective way to integrate science and language arts. Students are to record their observations, thoughts, and questions about past science experiences in journals to be kept in the science area. The observations may be recorded in sentences or sketches which keep track of changes both in the science item or in the thoughts and discussions of the students.

Magnetism and Electricity Journal entries can be completed as a team effort or an individual activity. Be sure to model the making and recording of observations several times when introducing the journals to the science area.

Use the student recordings in the Magnetism and Electricity Journals as a focus for class science discussions. You should lead these discussions and guide students with probing questions, but it is usually not necessary for you to give any explanation. Students come to accurate conclusions as a result of classmates' comments and your questioning. Magnetism and Electricity Journals can also become part of the students' portfolios and overall assessment program. Journals are a valuable assessment tool for parent and student conferences as well.

How To Make a Magnetism and Electricity Journal

1. Cut two pieces of 8.5" x 11" (22 cm x 28 cm) construction paper to create a cover. Reproduce page 83 and glue it to the front cover of the journal. Allow students to draw magnetism and electricity pictures in the box on the cover.
2. Insert several Magnetism and Electricity Journal pages. (See page 84.)
3. Staple together and cover stapled edge with book tape.

My Magnetism and Electricity Journal

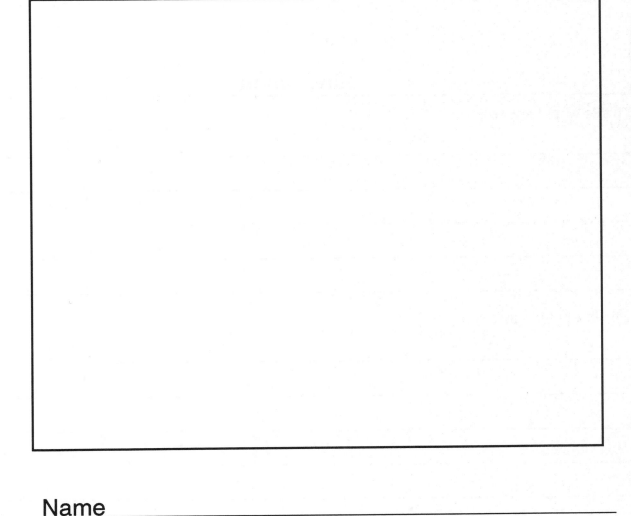

Name_____

Magnetism and Electricity Journal

Illustration

This is what happened:

_____

_____

_____

_____

_____

This is what I learned:

_____

_____

_____

_____

_____

My Science Activity

K-W-L Strategy

Answer each question about the topic you have chosen.

Topic:_____

K - What I Already **Know:** _____

_____

_____

_____

_____

W - What I **Want to Find Out:** _____

_____

_____

_____

_____

L - What I **Learned after Doing the Activity:** _____

_____

_____

_____

_____

Investigation Planner *(Option 1)*

Observation

_____

_____

Question

_____

_____

Hypothesis

_____

_____

Procedure

Materials Needed:

_____

_____

_____

Step-by-Step Directions: (Number each step!)

_____

_____

_____

_____

_____

Investigation Planner *(Option 2)*

Science Experience Form

Scientist _____

Title of Activity _____

Observation: What caused us to ask the question?

_____

_____

_____

Question: What do we want to find out?

_____

_____

_____

Hypothesis: What do we think we will find out?

_____

_____

_____

Procedure: How will we find out? (List step-by-step.)

1. _____

2. _____

3. _____

4. _____

Results: What actually happened?

_____

_____

_____

Conclusions: What did we learn?

_____

_____

_____

Magnetism and Electricity Observation Area

In addition to station-to-station activities, students should be given other opportunities for real-life science experiences. For example, small electric motors, circuit breakers, and simple crystal radio sets can provide vehicles for discovery learning if students are given time and space to observe them.

Set up a magnetism and electricity observation area in your classroom. As children visit this area during open work time, expect to hear stimulating conversations and questions among them. Encourage their curiosity but respect their independence!

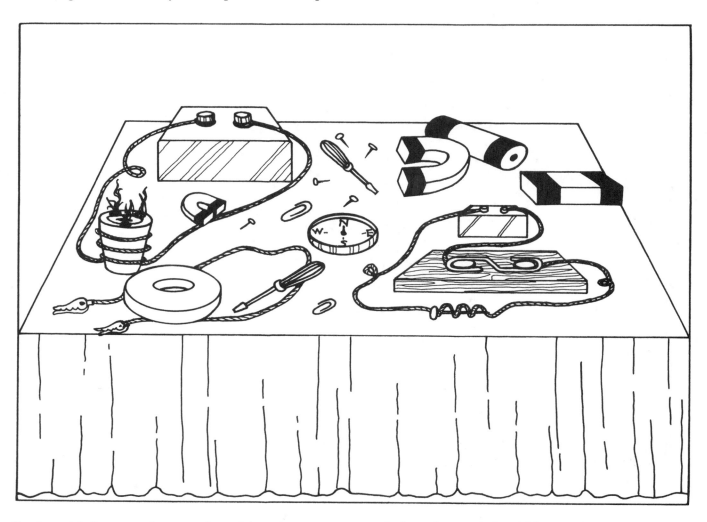

Books with facts pertinent to the subject, item, or process being observed should be provided for students who are ready to research more sophisticated information.

Sometimes it is very stimulating to set up a science experience or add something interesting to the Magnetism and Electricity Observation Area without a comment from you at all! If the experiment or materials in the observation area should not be disturbed, reinforce with students the need to observe without touching or picking up.

Assessment Form

The following chart can be used by the teacher to rate cooperative learning groups in a variety of settings.

Science Groups Evaluation Sheet

Room: _____   Date: _____

Activity: _____

Everyone

	Group									
	1	**2**	**3**	**4**	**5**	**6**	**7**	**8**	**9**	**10**
. . . gets started.										
. . . participates.										
. . . knows jobs.										
. . . solves group problems.										
. . . cooperates.										
. . . keeps noise down.										
. . . encourages others.										

Teacher comment

_____

Bragging rights for the group session: _____

_____

Assessment Form *(cont.)*

The evaluation form below provides student groups with the opportunity to evaluate the group's overall success.

Cooperative Group Evaluation

Assignment: _____

Date: _____

Scientists	**Jobs**
_____	_____
_____	_____
_____	_____
_____	_____

As a group, decide which face you should fill in and complete the remaining sentences.

1. We finished our assignment on time, and we did a good job.

2. We encouraged each other, and we cooperated with each other.

3. We did best at_____

_____.

4. Next time we could improve at _____

_____.

Assessment Form *(cont.)*

The following form may be used as part of the assessment process for hands-on science experiences.

Science Anecdotal Record Form

Date: _____

Scientist's Name:_____

Topic: _____

Assessment Situation: _____

Instructional Task: _____

Behavior/Skill Observed: _____

_____

_____

_____

_____

_____

_____

This behavior/skill is important because_____

_____

_____

_____

_____

_____.

Super Physicist Award

This is to certify that

_____
Name

made a science discovery.

Congratulations!

_____
Teacher

_____
Date

Glossary

A

Alternating Current—electric current that is constantly changing its direction of flow.

Ammeter—a special meter used for measuring electric current.

Anode—an electrode that is positively charged.

Atom—the smallest part of a substance. Inside the atom you will find the electrons, neutrons, and protons.

Attraction—when two charges or poles are different they "attract" or come together.

B

Battery—an object that produces an electric charge by means of a chemical reaction.

C

Capacitor—an electric component that stores an electrical charge.

Cathode—an electrode that is negatively charged.

Circuit—a closed path for electron flow.

Component—the part or piece of a circuit.

Conductor—material that allows the free flow of electrons, creating electric current.

Conclusion—the outcome of an investigation.

Control—a standard measure of comparison in an experiment. The control always stays constant.

Current Electricity—the continuous flow of electrons through a conductor.

D

Direct Current—the electric current flows in only one direction around the circuit.

E

Electric Current—a continuously flowing current of electric charges, either negative or positive.

Electricity—the flow of electrons through a conductor.

Electrode—a solid part of the battery, usually made from materials such as carbon and zinc.

Electromagnetism—an area of physics that studies the relationship between electricity and magnetism.

Electromagnet—a temporary magnet formed when an electric current flows through a wire coil. The coil is usually wrapped around an iron bar.

Electron—a particle inside an atom that carries a negative electric charge.

Electrostatic Attraction—the attraction that opposite electrical charges have for each other.

Experiment—a means of proving or disproving an hypothesis.

G

Geographic North—the earth's geographic north pole.

H

Hypothesis (hi-POTH-e-sis)—an educated guess to a question which one is trying to answer.

I

Insulator—material that will not allow electric current to flow through it.

Investigation—observation of something followed by a systematic inquiry in order to understand what was originally observed.

Glossary (cont.)

M

Magnet—an object that has a magnetic field around it.

Magnetic Induction—the process by which a substance (such as iron or steel) becomes magnetized by a magnetic field.

Magnetic Field—the area around a magnet that causes magnetic movement.

Magnetic North—the earth's magnetic north pole. This pole continually changes with the earth's magnetic field.

Magnetic Pole—the ends of a magnet. One pole is north, and one pole is south.

Magnetism—an invisible force that can make objects move away, move together, or stay in the same place.

N

Neutron—a particle inside an atom that carries a neutral charge.

O

Observation—careful notice or examination of something.

P

Parallel Circuit—this circuit provides more than one path for current; it also provides the same voltage for every source and output device.

Procedure—a series of steps that is carried out when doing an experiment.

Proton—a particle inside an atom that carries a positive electrical charge.

Q

Question—a formal way of inquiring about a particular topic.

R

Repel—to push away. When two charges or poles are the same, they *repel* each other.

Resistance—opposition to the flow of electrons.

Results—the data collected after performing an experiment.

S

Scientific Method—a creative and systematic process of proving or disproving a given question, following an observation. Observation, question, hypothesis, procedure, results, conclusion, and future investigations comprise the scientific method.

Scientific-Process Skills—the skills necessary to have in order to be able to think critically. Process skills include: observing, communicating, comparing, ordering, categorizing, relating, inferring, and applying.

Scientist—a person considered an expert in one or more areas of science.

Semi-Conductor—a material that conducts electricity better than insulators, but not as well as such conductors as copper.

Series Circuit—this circuit uses a single path to connect the electric source or sources to the ouptut device or devices.

Static Electricity—electrons that do not move but still create an electrical charge.

T

Transformer—a device that changes the voltage of electricity.

V

Variable—the changing factor of an experiment.

Volt—the unit of force by which electricity is measured.

Voltage—a force that causes electrons to flow.

Bibliography

Aaseng, Nathan. *The Inventors: Nobel Prizes in Chemistry, Physics, and Medicine.* Lerner, 1988.

Arco. *Energy, Forces, and Resources.* Arco, 1984.

Ardley, Neil. *Discovering Electricity.* Watts, 1984.

    Exploring Magnetism. Watts, 1984.

Asimov, Issac. *Asimov's Biographical Encyclopedia of Science and Technology.* Doubleday, 1982.

    How Did We Find Out About Superconductivity? Walker LB, 1988.

Boltz, C. L. *How Is Electricity Made?* Facts On File, 1985.

Bronowski, J, and Millicent E. Selsam. *Biography of an Atom.* HarperCollins, 1987.

Cobb, Vicki. *More Power To You!* Little Brown, 1986.

Cosner, Sharon. *The Light Bulb.* Walker & Co., 1984.

Cousins, Margaret. *The Story of Thomas Alva Edison.* Random House, 1981.

Fricke, Pam. *Careers with an Electric Company.* Lerner, 1984.

Gosnell, Kathee. *Electricity.* Teacher Created Materials, 1994.

Gutnik, Martin. *Electricity: From Faraday to Solar Generators.* Franklin Watts, 1986.

    Simple Electrical Devices. Watts, 1986.

Halacy, Dan. *Nuclear Energy.* Watts, 1984.

Hoyt, Marie. *Magnet Magic, Etc.* Educ Serv Pr., 1983.

Int Bk Center Staff. *Magnets, Bulbs, Batteries.* Intl Bk Ctr., 1987.

Jennings Terry. *Magnets.* Watts, 1990

Keith, Hal. *More Wires & Watts: Understanding and Using Electricity.* Scribner, 1981.

Lafferty, Peter. *Magnets to Generators: Projects with Magnetism.* Watts, 1989.

Lambert, Mark. *Future Sources of Energy.* Watts, 1986.

Lampton, Christopher. *Thomas Alva Edison.* Grey Castle Press, 1991.

Math, Irwin. *Wires and Watts: Understanding and Using Electricity.* Macmillan, 1981.

McKie, Robin. *Energy.* Hampstead, 1989.

    Solar Power. Gloucester, 1985.

Quackenbush, Robert. *Quick, Annie, Give Me a Catchy Line! Story of Samuel F.B. Morse.* Prentice Hall, 1983.

Richards, Norman. *Dreamers and Doers: Inventors Who Changed the World.* Macmillan, 1984.

Roberts, Royston M. *Serendipity—Accidental Discoveries in Science.* John Wiley & Sons, Inc., 1989.

Sabin, Louis. *Thomas Alva Edison: Young Inventor.* Troll, 1983.

Bibliography *(cont.)*

Santry, Laurence. *Magnets.* Troll Assocs, 1985.

Shippen, Katherine B. *Mr. Bell Invents the Telephone.* Random, 1963.

Strachan, James. *Future Sources.* Gloucester, 1985.

Vegland, Nancy. *Coils, Magnets, and Rings: Michael Faraday's World.* Coward, 1976.

Ward, Alan. *Magnets & Electricity.* Watts, 1992.

Weilbacher Mike. *Magnetism Exploration Kit: Discover One of Nature's Most Astonishing Forces.* Running Press, 1993.

Weiss, Ken. *How To Be an Inventor.* HarperCollins, 1980.

Whyman, Kathryn. *Electricity & Magnetism.* Gloucester, 1986.

Sparks to Power Station. Gloucester, 1989.

Wood, Robert W. *Physics for Kids.* Tab Books, 1990.

Spanish Titles

Biesty, S. *Del interior de las cosas (Incredible Cross-Sections).* Santillana Pub. Co., 1993.

Taylor, K. *Luz (Light).* Lectorum, 1992.

Technology

Agency for Instructional Technology. *Magnetism: Why Does a Compass Point North?* and *Electricity: Where Does Electricity Come From?* Available from AIT, (800)457-4509. video

Bill Walker Productions. *Electricity and Magnets.* Available from Cornet/MTI, (800)777-8100. film, video, and videodisc

D.C. Heath. *Electric Circuits* and *Electricity and Magnetism.* Available from William K. Bradford Pub. Co., (800)421-2009. software

Disney Educational Productions. *Electricity.* Available from Cornet/MTI, (800)777-8100. film, video, and videodisc

January Productions. *Electricity.* Available from CDL Software Shop, (800)637-0047. software

Microcomputer Workshop. *Making Circuits.* Available from CDL Software Shop, (800)637-0047. software

MicroMedia. *Basic Electricity.* Available from CDL Software Shop, (800)637-0047. software

Science Books & Films American Association for The Advancement of Science. *About Electricity.* Available from AIMS Media, (800)367-2467. videodisc

WDCN, Nashville. *Electricity.* Available from AIT, (800)457-4509. video